BULLETS DON'T FLY

Also by David Kim

The Real Money Guide A practical guide to financial recovery and building lasting wealth, drawn from the author's own journey through bankruptcy and back.

The Couples Money Guide (forthcoming) A guide to building financial partnership in marriage and long-term relationships.

The Kids Money Guide (forthcoming) Financial foundations for ages 12–18, preparing the next generation to avoid the mistakes and build the habits that matter.

BULLETS DON'T FLY

A Supply Marine's Memoir

DAVID KIM

ISBN:
9798992533033 (Ebook)
9798992533040 Hardback)
9798992533057 (Paperback)
9798992533064 (Audiobook)

Publisher: Self Published

Printed in the United States of America

First Edition

DEDICATION

To the unsung heroes of the Marine Corps, the men and women who serve in support roles, far from the front lines. To my fellow supply Marines, past and present, who understand the quiet strength it takes to keep the gears turning and the steel flying. This is for you.

To my family and friends, whose patience and love carried me through the toughest times. Their belief in me, even when I doubted myself, gave me the strength to find my way.

And to those who have struggled with the unseen wounds of service, the silent battles fought long after the uniform is shed. May this story offer understanding, shared experience, and a reminder that you are not alone.

CONTENTS

INTRODUCTION

The image of a Marine conjures battlefield heroism: charging across beaches, precision strikes from the air. These are powerful narratives, but they don't tell the whole story.

My battles weren't fought with rifles. They were fought on a PC, with paperwork, deadlines, and the endless task of ensuring our units had what they needed to function. I was a 3043, a supply Marine, and this memoir is an attempt to capture that often-unseen side military life.

Bullets Don't Fly follows my path from recruit training at MCRD San Diego through deployments to Kuwait, Saudi Arabia, and Korea, then into the civilian world of software engineering and consulting. It's about the mundane realities and unexpected adventures of a support role. It's about the people: the fellow Marines who became brothers, the drill instructors who shaped me, the loved ones who stood by me. It's about what happens when you take off the uniform and have to figure out who you are without it.

This is not a glorification of war. It's an honest portrayal of one Marine's experience: the humor and heartache, the challenges and growth, and the lasting impact of service on everything that follows.

THE MAKING OF A MARINE

Before the Marine Corps, there was Riverside. A kid who didn't fit in his own family, who stole from stores because he didn't know what else to do with the anger, who saw a recruiter's office as the only unlocked exit. I wasn't running toward anything. I was running away from everything.

The Corps caught me. Held me. Broke me down in boot camp and rebuilt me into something I didn't know I could be. Over eight years and three deployments, I learned that discipline wasn't punishment. It was freedom. That systems and logistics and supply chains weren't boring. They were the invisible architecture that made everything else possible. That I could be competent, respected, valuable.

Part I is the story of that making. From Riverside to Korea, from lost kid to sergeant, from escape to purpose. The foundation that everything after, the success, the failure, the rebuilding, would be built on.

CHAPTER 1

RIVERSIDE

101501ZAUG74

The path to the Marine Corps recruiting office in Riverside, California didn't start with a poster of dress blues or a recruiter's promise. It started decades earlier, on the other side of the world, with a woman who left everything she knew for a country that existed only as an idea. It wound through a Philadelphia hospital where doctors gave a premature infant a twenty percent chance of survival. It passed through Korean language schools and restaurant kitchens, through an auto shop that smelled of oil and possibility, through family arguments conducted in two languages and a grandmother who smuggled sandwiches to a starving boy. By the time I walked through that recruiting office door in 1991, I had already been shaped by forces I was only beginning to understand. This is where it began.

3

AGAINST THE ODDS

I wasn't supposed to survive.

Philadelphia, August 1974. I arrived ten weeks early, a premature infant in an era when neonatal medicine was still primitive by today's standards. The doctors gave me a twenty percent chance of survival. Maybe less. My mother faced an impossible choice: her life or mine. Her blood pressure had spiked to dangerous levels, a condition triggered by stress and abandonment, and the delivery was killing her.

She chose me.

The story of how I came to exist is inseparable from the story of my mother's courage and my biological father's cowardice.

My mother immigrated to the United States in the early 1970s through a nursing program, one of the pathways that brought skilled Korean women to American hospitals during a nationwide nursing shortage. She was ambitious, educated, determined to build a life in a country that promised opportunity for those willing to work for it. She met my biological father, married him, and became pregnant.

He abandoned her when she was three months along.

The marriage, I learned later, had been a transaction, at least for him. He used her to fast-track his residency status and his enlistment in the Air Force. Once he had what he needed, he disappeared. I've never seen him. Never spoken to him. Never received a letter, a phone call, an explanation. He exists only as an absence, a void in the shape of a father.

The stress of his desertion sent my mother's blood pressure into crisis. The premature delivery followed. And there I was: a tiny, fragile thing with a twenty percent chance of making it through the week.

I made it.

My mother brought my grandmother over from Korea to help raise me while she continued working as a nurse. This arrangement, three generations of Korean women in a Philadelphia apartment, became the foundation of my earliest years.

What my mother accomplished in that decade still astonishes me. Through the 1970s and into the early 1980s, she earned a Master's degree in Nursing while working full-time and raising a child as a single mother. She climbed from staff nurse to floor supervisor, a leadership role that was remarkable for anyone and nearly unheard of for a Korean immigrant woman in that era. The glass ceilings she broke, the prejudices she navigated, the exhaustion she must have carried, I understood none of it as a child. I only knew that we had a nice house, that my mother worked long hours, and that my grandmother was always there when I came home.

We had a good life. Respectable earnings for that time. Stability. A path forward.

Then my mother decided I needed a father.

The Move West

In the early 1980s, everything changed.

My mother had met a man, a Korean immigrant like herself, and convinced herself that I needed a father figure in my life. Maybe she believed it. Maybe she was lonely. Maybe the weight of single motherhood had finally become too heavy. Whatever her reasons, she made a decision that would redirect the course of both our lives.

We moved from Pennsylvania to California so she could marry my stepfather.

He had ideas about how things should be. A nursing career, he argued, was no life for a wife and mother. Too many hours. Too much independence, perhaps. He convinced her to abandon the profession she'd spent a decade mastering, the career that had carried us from immigration to stability, and to run restaurants with him instead.

I've thought about this decision countless times over the years. A Korean immigrant woman who had earned a Master's degree, who had risen to supervisor in a competitive hospital system, who had built a life through education and professional excellence, walking away from all of it to work in a restaurant kitchen. It had to be one of the worst decisions I could imagine. But that was the time. That was the culture. A wife followed her husband's vision, even when that vision meant trading a nursing career for a pizza oven.

I took my stepfather's surname. David Kim became David Yoon, at least in the classroom and in daily life. But my stepfather never formally adopted me. My birth certificate, my Social Security card, all my legal documents still said Kim. For ten years, I existed as two people: David Yoon at school, David Kim on paper. Nobody seemed to notice or care. In 1980s California, identity verification was loose enough that I even got a driver's license as David Yoon with no supporting paperwork.

I didn't think much about the duality until the day I tried to enlist. But that came later.

The Outsider

My childhood was shaped by two distinct forces: the deeply rooted traditions of my first-generation Korean immigrant family and the vibrant, often bewildering, culture of my American surroundings. These forces clashed constantly.

Growing up, the chasm between my family's customs and the norms of my predominantly white neighborhood and schools was vast and isolating. I was a constant outsider, acutely aware of the differences that set me apart. In a sea of familiar faces, I was often the only Asian student in my classes, a solitary figure even in high school, where I was one of only three Korean students among a student body exceeding two thousand.

The isolation ran deeper than culture. A boy without a biological father, carrying a name that didn't match his documents, living in a household where his mother had sacrificed her career for a man who would never fully accept him as a son. The weight of all these fractures pressed down constantly, even when I couldn't articulate what I was feeling.

This isolation shaped me. My introverted nature flourished, fed by a sense of otherness that permeated my school experience. Academically, I struggled, finding myself at odds with most of my teachers. Their methods, their expectations, simply didn't resonate with me, leading to strained relationships and less-than-stellar grades. Teachers accused me of cheating on exams more than once, forcing me to retake new versions alone in front of them. I'd score even higher the second time.

THE TROUBLED KID

To say I was a troubled kid with authority issues is an understatement.

Before high school, I'd had at least three run-ins with police, one arrest, and more runaway attempts than I care to count. The cultural collision between my parents' expectations and the reality of being a Korean kid in Riverside created a pressure cooker that periodically exploded in ways that satisfied no one.

The shoplifting incident stands out, not because it was my worst offense, but because of what happened afterward.

I was twelve years old, and 1986 was not a good year. I'd been caught stealing from a store, something stupid, something I didn't even need. The kind of petty theft that screams for attention more than acquisition. The cops charged me, released me to my parents, and a judge sentenced me to community service with mandatory counseling. The legal system had done its part. Now it was my parents' turn.

They beat the everliving shit out of me.

This wasn't unusual for Asian parents at the time; corporal punishment was simply how discipline worked in our household and in most Korean families we knew. But they didn't stop there. After the beating, they locked me in my room. For three days. No food. No water.

I'm still not sure this was legal. At the time, I thought I'd be better off in jail. At least they fed you there.

My Grandmother and the Hoagie

Our apartment on Olivewood Avenue in Riverside was small, two bedrooms that housed more people than the space was designed for. I shared one of those bedrooms with my grandmother, a woman who had survived more than I could comprehend at twelve years old. She'd been born in 1903 in Korea, lived through two world wars, the Japanese occupation from 1910 to 1945, and the Korean War. She'd borne ten children, though only two survived to adulthood, my mother, the youngest, born in 1944, and my aunt, the oldest, born in 1924.

Photo 1 My grandmother, Cha Yeon Kim
(김자연) She survived the Japanese occupation,
crossed an ocean, and still found ways to give.

My parents told my grandmother not to feed me while they were away during the day. She had one job: enforce my punishment. Let me suffer the consequences of my actions.

On the second day, maybe the third (the hunger had made time slippery), my grandmother walked into our shared room. She yelled at me. Called me an asshole in Korean. Her eyes were wet with tears, and I couldn't tell if she was crying because she was angry or because she was about to break the rules.

Then she reached into her pants and pulled out a footlong hoagie roll with ham and butter in the middle.

She'd smuggled it in. Hidden it in her clothing like contraband, which I suppose it was. After at least two days of starving, it was the most amazing thing I'd ever eaten. The bread was soft, the ham salty, the butter rich against my cracked lips. I devoured it in minutes.

She watched me eat, still crying, still angry. When I finished, she walked out, calling me an asshole again as she shut the door behind her.

That sandwich, that act of defiant love wrapped in deli meat, taught me something about the difference between rules and righteousness. My grandmother followed my parents' rules until following them became wrong. Then she found a way to do what was right while maintaining the appearance of compliance. It was a lesson in survival, in working within systems that don't serve you, in the small rebellions that preserve humanity within harsh structures.

To this day, I'm obsessed with French jambon-beurre sandwiches. Ham and butter on a baguette. Simple. Perfect. Every time I eat one, I think of her.

My grandmother lived until 2000, long enough to see both my children, her great-grandchildren, born. She died at ninety-seven years old, having witnessed the entire twentieth century from a perspective most Americans can't imagine. When I think about resilience, about surviving impossible circumstances with dignity intact, I think of her pulling that sandwich from her pants, tears streaming down her face, telling me I was an asshole in a language I barely spoke.

She was right, of course. I was an asshole. But I was her asshole. And she wasn't going to let me starve.

The Rich Aunt and the Commodore 64

My aunt's story was another matter entirely.

Born in 1924, she was twenty years older than my mother, practically a different generation. During the Japanese occupation of Korea, a Japanese family took her in her late teens to serve as a servant or nanny. The term varied depending on who told the story. My family claimed it was a job, not servitude, because they paid her a wage "secretly" into an account.

I've always had my doubts. It sounds more like they paid her off to avoid war crime charges than actual employment. But whatever the arrangement, my aunt took the money and made something of herself. She didn't return to Korea until well after the occupation ended, but when she did, she opened her own restaurant. She became successful. Wealthy, even.

Throughout my childhood, the "rich single aunt" would visit the United States once a year, arriving with gifts of clothing and money that made her seem almost mythical. She was everything my parents weren't, independent, prosperous, unburdened by the daily grind of running restaurants and raising a difficult child.

In 1987, my aunt changed my life. She bought me my first computer.

It was a Commodore 64, complete with monitor, printer, and disk drive, the whole kit. For a kid who'd found auto shop more interesting than academics, who felt perpetually out of place in school, that computer opened a universe I didn't know existed.

I taught myself programming, starting with BASIC and moving into whatever else I could find documentation for. I discovered piracy, the gray market of copied software that circulated among computer enthusiasts. And I found MUDs, those text-based multiplayer games that were the precursors to everything that would come later in online gaming.

The Commodore 64 was more than a machine. It was proof that my brain worked, just not in the ways school had been testing. I could build things, solve problems, create systems that functioned according to rules I understood. The logical architecture of programming made sense in ways that social dynamics never had.

My aunt permanently moved to the US in 1991, just one year before I joined the Marine Corps. The Commodore 64 had been replaced by a PC by then. My interest in computers had solidified into something that would eventually become a career. But I never forgot that initial gift, the moment someone saw potential in me and invested in it.

She was a complicated figure, my aunt. A survivor of occupation, a woman who turned victimhood into independence, who came from nothing and built something. Whether the money that funded her success was wages or hush payment, she'd taken what she had and made a life. There's something to respect in that, even if the origins were murky.

When I think about the women who shaped me, my grandmother with her smuggled sandwich, my aunt with her Commodore 64, I see a pattern. They were survivors who found ways to give, even when giving cost them something. My grandmother risked my parents' wrath. My aunt spent money she'd earned through circumstances she never fully explained.

Both of them saw something in me worth saving.

YOUNG'S PIZZA AND RIBS

The restaurant business was in my blood whether I wanted it there or not.

After my mother and I moved from Philadelphia, Pennsylvania to Upland, California in 1982, she and my new stepfather cycled through several establishments. First came Chopsticks, a Chinese restaurant in Upland. Then a sandwich shop in Los Angeles. Eventually, they landed on an Italian place in Riverside called Elliott's Pizza and Ribs, though my stepfather decided to change the name because it shared a strip mall with Elliott's Pet Emporium, and he thought having the same name as a pet store was weird.

So Elliott's became Young's Pizza and Ribs, and that's where I spent most of my childhood.

The menu was an interpretation of American comfort food: hand-tossed pizza baked in a proper Blodgett oven, Kansas City-style BBQ ribs and chicken that rotisseried all day, and Italian subs that were actually closer to good Jersey hoagies. The ribs were the real draw, slow-cooked until the meat fell off the bone. People came for the pizza but came back for those ribs.

I was free child labor that didn't fall under labor laws since I was a family member. Some days I worked from 8 AM setup to 10 PM cleanup. On school days, I walked from Magnolia Elementary or Central Middle to the restaurant, did my homework at one of the four square tables with their red-checkered plastic tablecloths, and then started working.

By ten years old, I'd tossed my first pizza round. I knew how to make the dough from scratch, the precise ratio of flour to water, the right amount of yeast, the kneading technique that made it elastic without being tough. I could prepare the rotisserie chicken, truss it properly, get it spinning on the spit at the right speed. I made all the sides from scratch and could build a Jersey sub blindfolded.

The only thing they didn't let me do was operate the big Hobart slicer, the one that sliced all the fresh vegetables and deli meats. Both my parents had stories of slicing parts of their fingers off on that very slicer. I was never sure if keeping me away from it was out of care for my safety or simply not wanting to lose productivity. Probably both.

The restaurant shaped my tastes in ways I didn't recognize until much later. To this day, I drink Diet Pepsi. Not because I prefer it, but because that was the least popular soda on our fountain, and sometimes syrup would go to waste from lack of sales. Waste was a sunk cost, and sunk costs were unacceptable. So I was allowed to drink as much Diet Pepsi as I wanted. It became habit, then preference.

Same with mortadella. That pink, pistachio-studded Italian deli meat was the least popular option for sandwiches, which meant I could eat as much of it as I wanted without cutting into profits. True Italian mortadella with pistachios is an art form, nothing like the yucky bologna Americans think it resembles. I still go out of my way to find the real thing.

My parents eventually sold Young's Pizza and moved on to other ventures: Sub King in Signal Hill, Volcano Burgers in Los Alamitos. Both were struggling restaurants they bought at a discount, then "flipped" like real estate, improving the menu and operations, turning them around, selling them for profit. They're both still operating today with 4.8+ Yelp ratings, long after my parents moved on.

The restaurant business taught me more than any classroom ever did. I learned operations, customer service, inventory management, the brutal mathematics of food cost versus pricing. I learned that success comes from showing up every day and doing the work, even when you're tired, even when you'd rather be anywhere else. I learned that a business is a living thing that requires constant attention.

And I learned about family expectations. In my parents' world, children contributed. There was no such thing as "quality time" that didn't involve productivity. Love was expressed through labor, through teaching skills, through preparing children for a world that didn't give handouts. Whether that approach was right or wrong, healthy or damaging, it was what I knew.

When I finally escaped to the Marine Corps, part of what I was escaping was the restaurant. The smell of pizza dough and BBQ smoke. The endless cycle of prep and service and cleanup. The feeling that my life had already been decided for me, that I would spend my years behind a counter in a strip mall, building someone else's dream.

FIRST LOVE

In the midst of cultural tensions and family pressures, Jenny showed up. She'd recently moved to Riverside from Rowland Heights, a slim Korean girl with flowing black hair who loved to dance. We started dating junior year, exchanging daily love letters filled with the kind of intensity only teenagers can muster.

These letters became a private sanctuary where we could say things we'd never say out loud. I kept them carefully hidden, a stash of emotion that belonged entirely to me in a household where my parents routinely disregarded privacy in the name of parental authority.

My mother found the letters while cleaning my room, though I always suspected she'd been searching for them. Finding them felt like a betrayal. The contents, with their references to physical intimacy and teenage passion, collided head-on with her religious conservatism.

"이런 편지가 뭐야?" (What are these letters?) she had demanded, waving the stack of papers like evidence in a trial. "This is sin! This is not proper behavior!" Her religious convictions and cultural expectations collided in a torrent of criticism, demanding that I end the relationship immediately.

That night, I did something I had never done before, I stood up to my mother, defending Jenny and my right to feel, to love, to make my own choices. The argument raged for hours, my mother invoking God, Korean values, and parental authority, while I fought for something I barely understood but felt deeply, the right to my own heart.

My mother turned out to be right, though not in the way she'd predicted. Early in my senior year, Jenny broke up with me and started dating a close friend within weeks. The betrayal cut deep. Looking back, this heartbreak probably pushed me toward the Marine Corps, a choice that promised escape from family pressures and the wreckage of first love.

The end of the relationship left me adrift. The daily ritual of letter writing had become such a part of my life that its absence felt like a physical void. The hallways at Poly became a minefield of awkward encounters and avoided glances, made worse by watching Jenny and my former friend together. The auto shop became even more of a sanctuary during this period, the precise demands of mechanical work offering a refuge from emotional chaos.

The breakup happened just as things at home were getting worse. Where I'd once found escape in romance, now I needed something more drastic. The Marine Corps recruiting office, with its promises of independence and reinvention, offered a complete break from both family constraints and painful memories.

MR. REDWINE'S SHOP

Mr. Redwine's auto shop occupied the far corner of Riverside Poly's campus, a cavernous space that smelled of motor oil, metal, and possibility. The building itself was unremarkable, standard-issue industrial architecture with roll-up doors and concrete floors permanently stained with decades of automotive fluids. But from the moment I first walked through those doors my sophomore year, I sensed something different about this space. Here, the complex social hierarchies and cultural expectations that dominated the rest of my life seemed to evaporate.

The shop operated on its own rhythm, distinct from the rigid schedules that governed the rest of the school. While other classrooms enforced strict silence and orderly rows of desks, the auto shop hummed with purposeful noise: the metallic clang of tools, the whir of pneumatic wrenches, the low rumble of engines coming to life. These sounds became my favorite kind of music, a language of mechanical precision that required no cultural translation.

The physical space showed its history. Tool chests lined the walls, each drawer meticulously organized, each tool bearing the marks of years of use. Lifts occupied the center of the shop, their hydraulic arms raising vehicles to expose the mechanical mysteries beneath. The air carried a permanent haze of oil mist and metal dust, a tangible reminder of the real work that happened here.

I spent my first few weeks in auto shop hanging back, watching how other students approached their projects, studying the way they handled tools and diagnosed problems. Mr. Redwine noticed my cautious approach but didn't push. Instead, he would casually drift by my workstation, offering small tips or asking quietly probing questions about what I observed. His teaching style was subtle but deliberate, allowing students to discover their own path while providing gentle guidance.

"A good mechanic," he told me one afternoon, "learns more from watching and listening than from rushing in. You've got that part figured out already." His words carried a weight of validation I rarely experienced elsewhere. In the rest of my life, my tendency toward quiet observation was often seen as a deficit. Here, it was recognized as a strength.

My first real project was a carburetor rebuild. While other students rushed through their work, I approached mine with methodical precision. I carefully removed each component, cleaned it, and inspected it. I created detailed diagrams of the disassembly process, making notes about the position and condition of each part. What might have seemed like excessive caution to others felt natural to me; this was how my Korean grandmother approached the preparation of traditional dishes, with respect for each ingredient and step.

The parallel between mechanical work and my grandmother's cooking became increasingly clear. Both required patience, precision, and an understanding of how components worked together. Just as she could tell the readiness of a dish by its aroma or the way it moved in the pot, I began to develop an intuition for engines, learning to diagnose problems by sound, vibration, and smell. This connection between my cultural heritage and mechanical work provided an unexpected bridge between my divided worlds.

Mr. Redwine would often stop by my bench, examining my work without comment. One day, as I carefully measured the float height on a rebuilt carburetor, he broke his usual silence. "You know, David, there are two kinds of mechanics in this world. Some just want to get the job done. Others want to understand why things work the way they do. You're the second kind." His observation went beyond mere mechanical aptitude; it was an acknowledgment of a deeper approach to learning and problem-solving.

The shop became my sanctuary during lunch breaks and after school. While other students rushed to the cafeteria or sports fields, I would slip into the quiet shop, working on whatever project was available. Sometimes it was routine maintenance on the school's aged fleet of practice vehicles. Other times it was helping fellow students diagnose particularly stubborn problems with their cars. In these moments, the cultural complexities that defined the rest of my life faded away, replaced by the straightforward logic of mechanical systems.

The technical library in the corner of the shop became my personal retreat. The Chilton manuals, with their detailed diagrams and systematic procedures, became my favorite reading material. Unlike the literature we studied in English class, these books spoke a language I inherently understood. They were direct, logical, and free from the cultural nuances that often tripped me up in other subjects. A torque specification was a torque specification, regardless of who was turning the wrench.

Mr. Redwine's background as a former US Navy Sailor gradually emerged through casual conversations over engine blocks and transmission rebuilds. His military experience shaped his teaching style, demanding excellence while providing the tools and guidance to achieve it. He never mentioned my cultural background or family situation, yet somehow his teaching style bridged the gaps I felt in other parts of my life.

The other students in the shop came from diverse backgrounds: future mechanics, curious hobbyists, and some who just needed an elective credit. But in the shop, these differences mattered less than one's ability to solve problems and work carefully. I formed easy friendships over shared projects, our conversations focused on mechanical challenges rather than the social pressures that dominated life outside the shop. For the first time, I experienced what it felt like to be judged solely on the quality of my work rather than my cultural identity or family background.

The Malibu: Lessons in Transformation

The 1967 Chevrolet Malibu SS sat beneath layers of dust in Mr. Redwine's cousin's barn, its once-proud lines barely visible through years of neglect. It was early spring of 1991, and Mr. Redwine had brought me to see the car after school, driving us out to the edges of Riverside near Norco where development gave way to scattered farms and aging citrus groves. The drive felt significant, a departure from our usual teacher-student dynamic into something more meaningful.

"Most people would see this as junk," Mr. Redwine said, running his hand along the car's weathered canary yellow frame. His touch was almost reverent, like a doctor examining a patient. "But you know what I see, David? I see potential. Just like I see in you." He glanced at me with those keen eyes that seemed to look straight through my carefully maintained facade. "Want to prove them wrong?"

The negotiation for the car became my first lesson in real-world diplomacy. Mr. Redwine didn't just handle it himself; he walked me through the process, teaching me how to assess value beyond the surface, how to negotiate respectfully but firmly. The art of negotiation he demonstrated that day, a careful balance of confidence and humility - served me well in my military career, particularly in dealing with international suppliers and contractors.

When we finally settled on a price of $800, he made me count out the money myself, cash I'd saved from working at my parents' restaurant. "When you pay for something with your own money," he said, "you treat it differently. You invest yourself." The weight of those bills in my hand, representing many hours of work, gave the transaction a gravity that would shape my approach to the entire project.

The Malibu's journey from the barn to the school shop was an adventure in itself. We had to borrow a trailer, and the loading process took hours as we carefully worked around seized wheels and corroded parts. The whole time, Mr. Redwine shared stories from his Navy days, weaving lessons about teamwork and perseverance into tales of military life.

"In the Navy," he explained as we winched the car onto the trailer, "we learned that every big mission is just a series of small tasks done right. You don't restore a car in one day. You don't become a Marine in one day either. It's about showing up, doing the work, and trusting the process." His words carried the weight of experience, offering a philosophy that went far beyond automotive restoration.

The engine became our first major project. After carefully extracting it from the car, we discovered the full extent of the challenge. Years of sitting had locked the pistons in their cylinders, and moisture had corroded many of the internal components. Mr. Redwine turned this into a master class in problem-solving.

"Start with what you know," he guided me through the diagnosis. "Then work your way toward what you don't know. Just like in the military - you gather intelligence before you make a plan." We spent weeks just on the engine, disassembling it piece by piece, documenting every step, identifying what could be salvaged and what needed to be replaced. Each component told its own story of neglect and potential renewal.

The 350 small block V8 gradually emerged from its years of neglect under our careful attention. Mr. Redwine taught me the importance of precision, how a few thousandths of an inch in bearing clearance could mean the difference between a healthy engine and catastrophic failure. These lessons in mechanical precision offered a clarity I couldn't find in other parts of my life.

"Slow is smooth, smooth is fast," became his mantra during our work sessions. When frustration would build, when a stubborn bolt wouldn't break free or a part wouldn't fit quite right, he'd remind me to step back, think it through, and approach the problem methodically. This wasn't just about car restoration; it was about life itself.

Parts acquisition became another lesson in resourcefulness and networking. We scoured junkyards, negotiated with suppliers, and built relationships with other restoration enthusiasts. Mr. Redwine taught me how to handle these networks, how to build trust with suppliers, how to find rare parts through unconventional channels. These skills in supply chain management and relationship building helped in my later military logistics career.

The bodywork presented its own unique challenges. Rust had eaten through significant sections of the frame, requiring careful cutting and welding to maintain structural integrity. Mr. Redwine used these repairs to teach principles of structural engineering, explaining how forces moved through the car's frame and why each repair needed to be precisely executed. His lessons went beyond mechanical knowledge to include broader principles of system integrity and reliability.

"You see this rust?" he asked one day, pointing to a particularly corroded section of frame we were repairing. "Rust is like doubt, it eats away at you from the inside out. But if you catch it early, if you deal with it honestly, you can stop it from spreading. Same goes for problems in life."

As the months passed, the Malibu slowly transformed. Each step brought new challenges and new lessons. Mr. Redwine used every moment as a teaching opportunity, not just about mechanics, but about life, leadership, and personal growth. He shared stories from his time in Vietnam, not the typical war stories, but quieter tales about responsibility, about taking care of your people, about maintaining your integrity under pressure.

The electrical system proved especially challenging. Years of exposure had corroded wires and connections throughout the car. Rather than simply replacing the wiring harness, Mr. Redwine insisted we trace each circuit, understand its function, and repair or replace components methodically. This painstaking process taught me about the importance of attention to detail and the value of understanding systems holistically.

"Every wire has a purpose," he explained as we mapped the complex electrical pathways. "Just like every Sailor and Marine in a unit has a role. You can't just focus on the big components and ignore the smaller connections."

The other students in the shop noticed our project. Some offered to help, and Mr. Redwine used these moments to teach me about leadership: how to delegate tasks, how to teach others, how to build a team. Under his guidance, the Malibu project became a focal point for collaboration and learning. I fell naturally into a leadership role, coordinating different aspects of the restoration while ensuring we maintained quality standards.

The interior restoration became an exercise in patience and creativity. Weather and rodents had destroyed the original seats beyond salvation. We located replacement seats from a donor car, but they required significant modification to fit properly. Adapting and problem-solving through this process taught me valuable lessons about working with available resources and finding creative solutions to unexpected challenges.

"Sometimes," Mr. Redwine noted as we modified the seat brackets, "the perfect solution isn't available. That's when you have to get creative while maintaining standards. It's about finding the balance between innovation and reliability."

Weekend work sessions became regular events. I would arrive early on Saturday mornings, often finding Mr. Redwine already there, coffee in hand, ready to tackle the next phase of the project. These informal sessions allowed for deeper conversations about life, career choices, and future possibilities. It was during one of these morning sessions that he first seriously discussed the Marine Corps as a potential path for my future.

By the spring of my junior year, the Malibu was nearing completion. The engine purred like new, the body gleamed with fresh paint, and we had completely restored the interior. But the real change wasn't in the car; it was in me. Through this project, I had discovered a confidence I'd never known before, a sense of capability that transcended cultural barriers and family expectations.

The final assembly phase brought all our previous work together. Each component we had carefully restored now needed to be integrated into a functioning whole. This process of systems integration taught me valuable lessons about project management and the importance of seeing both details and the bigger picture. Mr. Redwine used this phase to reinforce earlier lessons about leadership and responsibility.

"Notice how everything has to work together," he observed as we made final adjustments to the timing. "One component out of sync can affect the entire system. That's true in mechanics, in the military, in life.

Success depends on understanding these relationships." His words carried extra weight now, backed by months of hands-on experience and visible results.

The day we fired up the engine for the final time was a moment of triumph and reflection. The sound of the V8 roaring to life represented the culmination of many hours of work, problem-solving, and personal growth. Mr. Redwine stood back, letting me take the lead in the final adjustments. As the engine settled into a smooth idle, his eyes glistened with a pride that went far beyond automotive achievement.

"You did this," he said simply, placing a heavy hand on my shoulder. "Remember that feeling. It's what happens when you believe in yourself as much as others believe in you." The moment was powerful in its simplicity.

The Malibu's restoration had become known throughout the school. Teachers and students who had previously seen me as just another quiet Asian student now saw something different: someone capable of taking on complex challenges and seeing them through to completion. This shift in how others saw me matched an internal shift; I saw myself and my possibilities differently now too.

The documentation we had maintained throughout the project became a comprehensive record of both technical achievement and personal growth. Looking through the project notebook, with its detailed photographs, diagrams, and notes, I could trace my development from hesitant beginner to confident project leader. Mr. Redwine encouraged me to use this documentation as a portfolio, understanding that it demonstrated real capability.

"This shows something important," he said, reviewing the notebook. "It shows you can take a complex project from start to finish. That's a rare ability, whether you're working on cars, leading Marines, or running a business. You've proven you can do it." His words helped me see how the skills I'd developed could translate into different career paths.

The first drive in the completed Malibu was a moment of pure joy and validation. Mr. Redwine insisted I take the wheel, watching with quiet satisfaction as I navigated the car through its first test drive. The smooth operation of every system we had restored, the solid feel of the controls, the deep rumble of the engine, all testified to the quality of our work and the value of doing things right.

"A project like this," Mr. Redwine reflected as we drove, "teaches you something about transformation. Sometimes you have to completely disassemble something, understand each part, before you can rebuild it better than it was. That's true for cars, for organizations, for people." His words carried a deeper meaning that would influence my approach to future challenges, both personal and professional.

BREAKING POINT

The arguments always started the same way.

Quiet comments about my future would escalate into explosive confrontations. Our small house near the 91 freeway would fill with raised voices in two languages, Korean and English colliding in a cacophony of cultural conflict. It was early 1992, and the tension that had been building for years was about to snap.

The success with the Malibu, rather than easing things, seemed to make them worse. While Mr. Redwine saw potential in my mechanical abilities, my parents viewed my growing passion for automotive work as a distraction from what they considered a proper academic path.

"대학! College!" My mother would begin, her voice carrying that particular tone of immigrant determination that had driven her from a small village in Korea to owning a restaurant in California. "You think working on cars is a future? You think this is what we came to America for?"

My stepfather would stand in the kitchen doorway, arms crossed. Despite being a first-generation immigrant himself, he had aligned completely with my mother's vision. "You live in this house, you follow our rules," he would declare. "American children have no respect. You want to be like them? Lazy? Useless?"

My mother's sacrifices became weapons. She would recount every hardship she had endured: the long hours at the restaurant, the humiliation of learning English as an adult, the countless small indignities of immigrant life. All of it presented as debt I needed to repay through academic achievement and a prestigious career.

The irony wasn't lost on me. They had left Korea seeking opportunity and freedom, yet now sought to restrict my own pursuit of the same. The courage that had brought them across an ocean somehow didn't extend to letting their son explore possibilities beyond their immediate control.

The final confrontation came on a Sunday evening. I had returned late from working on the Malibu, my hands still carrying traces of engine grease. My mother sat at the kitchen table, a stack of college brochures spread before her like evidence in a trial. Riverside Community College. Cal State San Bernardino. UC Riverside.

"We've made our decision," she announced. "You will attend college locally. You will live at home. Nine o'clock curfew will continue." She laid out each point like a contract. "This is not American home. This is Korean home. Our rules."

Every night, I watched them close the restaurant at nine, counting coins and organizing the next day's prep lists. Was this to be my future? Living under their roof, following their rules, attending a local college while working in the family business until I was old enough to take it over?

The contrast between my two worlds had never felt more stark. In the auto shop, I was trusted with complex projects, given responsibility, treated as capable. At home, I was still a child to be controlled, my achievements dismissed unless they aligned with their narrow vision of success.

"I need time to think," I said, retreating to my room.

My mother followed, continuing through the closed door. "You think Mr. Redwine cares about your future? You think fixing cars is a future? Doctor, lawyer, engineer—these are futures!" Her voice cracked, revealing the fear beneath her anger. "You want to struggle like us? Work twelve hours every day, no vacation, no security?"

That night, lying in bed, I could hear them talking in the kitchen, discussing my future as if my own desires were irrelevant. The nine o'clock curfew. The local college mandate. The suffocating control.

I realized that staying meant surrendering not just my independence but my identity. The respect I'd found in Mr. Redwine's shop, the confidence I'd built restoring the Malibu, the sense of self I'd begun to discover—all of it would be sacrificed on the altar of their expectations.

The Marine Corps recruiting office sat on a corner I passed every day on my way to help at the restaurant. I'd never given it much thought before. Now, the posters in the window showing proud Marines in dress blues spoke of a different kind of discipline—not the arbitrary rules of immigrant parents, but a structured path to genuine independence.

The next morning, instead of heading to first period, I pulled into the recruiting office parking lot.

The decision wasn't rebellion. It was survival. Just as I had learned to diagnose mechanical problems systematically in the auto shop, I had diagnosed the problem in my life: I needed a way to break free that would also prove my worth. A path that would demand more of me, not less.

I walked through the door.

The recruiting office door felt heavier than it looked. Inside, the air conditioning hit me like a wave of clarity, cutting through the heavy Riverside heat. The Marine at the front desk looked up, his bearing reminiscent of Mr. Redwine's quiet confidence. "Can I help you?" he asked, his voice carrying the kind of assured authority I'd only heard from Mr. Redwine. There was no judgment in his eyes, no immediate assumptions about the Asian kid standing in his office. Just professional interest and something else: recognition, perhaps, of the look every recruiter must know. The look of someone searching for a way out.

"What's the difference between the services?" I managed to ask, trying to keep my voice steady. "Other than the Marines having the best-looking uniform?" The attempt at humor felt weak, but it brought a knowing smile to the recruiter's face. This small interaction marked the beginning of what would become a life-changing conversation.

The question brought Gunnery Sergeant Reefer from his office. He moved with the kind of assured purpose I'd never seen in my stepfather, not the bluster of assumed authority, but the quiet confidence of earned respect. His office, when he invited me in, was a study in everything my home life wasn't: organized, purposeful, clear in its intentions.

"So," he said, settling behind his desk, "tell me what brought you here today."

The story poured out of me, not just the previous night's ultimatum, but everything. The cultural pressures, the suffocating rules, the nine o'clock curfew that felt like a chain around my neck. I told him about the restaurant, about watching my parents work twelve-hour days while demanding I follow a path, I couldn't see myself taking. About a stepfather who commanded respect without earning it, and a mother whose love felt like a cage.

GySgt Reefer listened with the kind of attention I'd only experienced in Mr. Redwine's shop, not just hearing my words but understanding the weight behind them. When I finished, he was quiet for a moment, studying me with experienced eyes.

"Let's start with the paperwork," he said, pulling out a folder. "Birth certificate, Social Security card, driver's license."

I handed them over and watched his expression shift. He spread the documents across his desk. Birth certificate: David Kim. Social Security card: David Kim. Driver's license: David Yoon.

"Which one are you?"

I explained: the biological father who'd abandoned my mother before I was born, using their marriage to fast-track his way into the Air Force and American citizenship. The stepfather who'd given me his name when I was eight but never formally adopted me. Ten years of being David Yoon in classrooms and on diplomas while remaining David Kim on every legal document. In 1980s California, nobody had questioned it. I'd even gotten that driver's license as Yoon with no supporting paperwork.

GySgt Reefer tapped my birth certificate. "Son, you're enlisting in the United States Marine Corps. You're going to be whoever this says you are." He looked up. "David Kim. That's your name now."

The irony wasn't lost on me. My biological father had used the military to escape his responsibilities. I was using the same institution to reclaim the identity he'd left behind.

"You know," he said finally, "most people come in here talking about serving their country, following family tradition, or looking for adventure." He leaned forward slightly. "But sometimes, the best Marines are the ones looking to prove something, not to others, but to themselves."

His next words struck a chord. "With your obvious intellect, you could choose any Military Occupational Specialty you want. Intelligence,

Administration, Supply, these roles require the kind of attention to detail I can already see in you."

The six-year contract he outlined wasn't just an enlistment agreement; it was a declaration of independence. The longer enlistment contract, than the standard four-year contract, included a guaranteed placement in one of three technical specialties and accelerated promotion to Lance Corporal (E-3). The Montgomery GI Bill would also put college within reach, but on my terms, not my parents'.

"But here's what you need to understand," he said, his tone growing serious. "The Marine Corps isn't an escape; it's a transformation. We're not interested in people running away from something. We want people running toward something greater than themselves."

The word caught me, reminding me of the Malibu sitting outside. Just as Mr. Redwine had helped me see the potential in that rusted shell of a car, GySgt Reefer was offering me the chance to see the potential in myself.

The Enlistment Reveal

My parents found out about my enlistment the way they found out about most things in my life: after the fact, when it was too late to change.

Gunny Reefer showed up at their restaurant.

I hadn't told my recruiter that my parents didn't know. I hadn't told him that I'd been planning this escape for months, waiting until I was seventeen and legally able to enlist with parental consent. I definitely hadn't told him that "parental consent" was going to be the hardest part of this entire process.

The Gunny walked into Young's Pizza and Ribs on a Tuesday afternoon, dress blues crisp, cover tucked under his arm. The restaurant was in its usual dinner rush when we arrived. The familiar smell of pizza and BBQ filled the air, but everything felt different now. I was no longer just the owner's son, I was someone making a choice that would reshape not just my future, but our entire family dynamic.

"Mrs. Kim? I'm here about your son David."

"What did he do?", my mother replied.

"Nothing wrong, ma'am. He wants to join the Marines. I just need you to sign some papers."

According to my mother's retelling—which I've heard approximately forty-seven times—she called my stepfather out of the kitchen, and they both stared at Gunny Reefer like he'd announced I was joining a circus. Which, in their minds, was probably an apt comparison.

The conversation that followed was conducted in rapid Korean, too fast for the Gunny to follow, punctuated by occasional gestures in his direction and what I can only assume were questions about what kind of family produces a son who makes life decisions without telling anyone.

Eventually, they signed.

I don't know what convinced them. Maybe they realized I'd find a way to leave regardless. Maybe they saw it as a path to discipline for a kid who'd already been arrested, run away, and generally failed to meet expectations. Maybe they were just tired.

When I got home that evening, my mother's only words were: "You live in our house, you follow our rules. You want to go? Go."

So I went.

The next morning, I walked into Mr. Redwine's shop. He looked up from the engine he was working on, and something in my bearing must have told him everything.

"You made your decision," he said. It wasn't a question.

"Yes, sir. I'm joining the Marines."

His smile was subtle but proud. "Good."

CHAPTER 2

BOOT CAMP

110030ZAUG92

The bus pulled into Marine Corps Recruit Depot San Diego thirty minutes before my eighteenth birthday ended. I didn't know it then, but the person who stepped off that bus would cease to exist over the next thirteen weeks. The Marine Corps has a term for what happens at boot camp: transformation. But that word is too clean, too clinical for what actually occurs. What happens is demolition followed by reconstruction. They break you down to your component parts, discard everything that doesn't serve the mission, and rebuild what remains into something harder, sharper, and no longer entirely your own. Eighty recruits stepped onto those yellow footprints that August night. Not all of us would make it to graduation. But those who did would carry something for the rest of our lives—a identity forged in sweat and sand and the constant voice of drill instructors who wanted us to fail so badly that we'd refuse to give them the satisfaction.

THE YELLOW FOOTPRINTS

The Greyhound bus pulled up to Marine Corps Recruit Depot San Diego at 0030 hours on August 11, 1992, my eighteenth birthday ending just 30 minutes prior. Through the smudged window, I could see the iconic yellow footprints painted on the concrete, glowing faintly under the sodium vapor lights. Those footprints had been the first step for generations of Marines before me. Now they waited for my own uncertain stride.

The bus's air conditioning had failed somewhere around Oceanside, leaving us to marinate in the collective anxiety of forty young men about to have their lives upended. The salty breeze from San Diego Bay carried the distant sound of aircraft from the nearby international airport, a final reminder of the civilian world we were about to leave behind.

My last glimpse of the civilian world through those windows was surreal, the lights of San Diego twinkling in the distance, palm trees swaying in the marine breeze, all of it about to be replaced by a reality I could barely imagine. The Malibu was safely stored in my parents' garage, covered and waiting, like a piece of my former life preserved in amber.

"Listen up, you have two minutes to get off my bus! Two minutes!" The driver's voice cracked through the heavy air, triggering a scramble of bodies and duffel bags. The organized departure we'd practiced at the Military Entrance Processing Station dissolved into chaos. Someone's bag caught on a seat; another recruit stumbled in the aisle. Two minutes felt like two seconds.

Then we heard it, a sound that would become burned into our memories. "GET OFF MY BUS! GET OFF! GET OFF! GET OFF!" The thunderous voice seemed to come from everywhere at once. A Marine Drill Instructor materialized at the bus door, his campaign cover casting sharp shadows across his face. His presence was electric, sending shockwaves of fear through our disorganized ranks.

"YOU HAVE JUST ARRIVED AT MARINE CORPS RECRUIT DEPOT SAN DIEGO! YOU ARE NOW ABOARD MY BELOVED

CORPS' RECRUIT TRAINING FACILITY! THE FIRST AND LAST WORDS OUT OF YOUR FILTHY MOUTHS WILL BE 'SIR'! DO YOU UNDERSTAND?"

"Sir, yes, sir!" The response was ragged, uncoordinated. Some recruits shouted, others barely whispered. I found myself somewhere in between, my voice caught in my throat like all the Korean honorifics I'd grown up struggling to master. This was a new language of respect, more demanding than any I'd known before.

"I CAN'T HEAR YOU, MAGGOTS!"

"SIR, YES, SIR!" This time, the response was stronger, born of pure terror and the sudden realization that half-measures would not be tolerated here. The collective voice of forty terrified recruits echoed off the depot's buildings, coming back to us like a promise we'd better be prepared to keep.

"STAND ON THE YELLOW FOOTPRINTS! MOVE!"

We poured out of the bus like water finding its level, each of us desperately seeking our place on those legendary footprints. I found mine near the middle of the formation, my feet automatically aligning with the painted outlines. The concrete was still warm from the day's heat, radiating through the thin soles of my sneakers, the last time I would wear civilian shoes for the next three months.

What followed was a blur of controlled chaos. More Drill Instructors emerged from the shadows, their voices creating a wall of coordinated intimidation. Their campaign covers, silhouetted against the night sky, looked like the beaks of predatory birds. One appeared inches from my face; his words precisely measured for maximum impact.

"YOU ARE NO LONGER CIVILIANS! YOU ARE NO LONGER ENTITLED TO THE RIGHTS AND PRIVILEGES OF CIVILIAN LIFE! YOU HAVE VOLUNTEERED TO BECOME PART OF THE FINEST FIGHTING FORCE IN THE HISTORY

OF MANKIND! BUT FIRST, YOU MUST PROVE YOU'RE WORTHY OF THAT TITLE!"

The speech continued, each word designed to strip away our civilian identities. They taught us the position of attention: heels together, feet at a 45-degree angle, thumbs along trouser seams, eyes straight ahead. Every movement had to be precise, every response immediate. The casual slouches and individual mannerisms we'd brought with us began to fade under the relentless pressure for uniformity.

Processing followed, a mechanized system designed to transform civilians into raw material for the Marine Corps. They herded us through a series of stations: medical screening, gear issue, administrative processing. Each step further dissolved our individual identities into a collective mass of potential Marines. The efficiency of the process was both impressive and terrifying, like watching a machine designed to disassemble human beings and rebuild them according to Marine Corps specifications.

The haircuts came next. I watched in the mirror as the clippers removed any trace of personal style, reducing each of us to identical blank canvases. My thick black hair fell away in clumps, joining the multicultural pile on the floor, all of us becoming the same under the buzzing blades. The barbers worked with mechanical efficiency, treating each head like an assembly line component that needed to be stripped and standardized.

Then came the phone call, one minute to contact our families. I stood in line, clutching the script we'd been given, watching as recruit after recruit delivered the same rehearsed messages home. When my turn came, the phone felt foreign in my hand. The line connected, and my mother's voice answered.

"Mom," I began, following the script but feeling the words catch in my throat. "I've arrived safely at Marine Corps Recruit Depot San Diego.

Please do not send any food or bulky items. I will contact you by mail with an address."

"David?" Her voice carried concern, confusion, "Are you-"

"I have to go now. I love you." The words tumbled out just as the Drill Instructor's hand reached for the phone. The connection cut off, severing my last link to the civilian world. In that brief exchange, I heard all the unspoken emotions in my mother's voice, the worry of an immigrant parent whose son had chosen an unfamiliar path, the pride she couldn't quite hide, the fear she tried to suppress.

The gear issue process was a blur of activity and shouted instructions. Each of us received identical sets of equipment, uniforms, boots, hygiene items, all precisely arranged according to military specifications. Every item had to be marked, stored, and maintained according to exact standards.

We stuffed our civilian clothes into boxes for storage. Watching my jeans and t-shirts disappear into that box felt like watching my old identity being packed away. The new uniform felt stiff and foreign against my skin, the boots heavy and unyielding.

That first night in the squad bay was a study in controlled discomfort. Sixty recruits arranged in perfect rows of racks, each of us learning the art of instant obedience. The squad bay itself was a massive room that seemed designed to strip away any sense of privacy or individuality. The overhead lights cast harsh shadows, making everything feel exposed and vulnerable.

"Sleep aye aye, sir!" we shouted in response to the command for lights out. But sleep was elusive. The thin mattress, the unfamiliar sounds, the lingering shock of our arrival, everything conspired to keep us awake. The night was punctuated by the steady footsteps of the fire watch, recruits assigned to patrol the squad bay in two-hour shifts.

Lying there in the dark, I thought about the yellow footprints outside. They had been my first step into this new world, a world stripping away individual identity to make room for something else: something stronger, more disciplined, more unified. My eighteenth birthday had marked not just a legal transition to adulthood, but the beginning of something I couldn't yet name.

The sounds of the squad bay filled the darkness: the collective breathing of sixty nervous recruits, the occasional muffled sob, the squeak of springs as someone shifted uncomfortably on their rack. Through the high windows, the glow of the base's lights cast weird shadows across the ceiling. Every now and then, the fire watch's flashlight beam would sweep across the room, a reminder that even in sleep, we were being monitored.

In the distance, I could hear the faint sound of another bus arriving, another group of civilians finding their places on those yellow footprints. The cycle was continuous, endless, the Marine Corps' perpetual process of breaking down and rebuilding, of making Marines. I was now part of that process.

Sleep came in fitful bursts, interrupted by the unfamiliar sounds and the anxiety of what was to come. Each time I drifted off, I was jolted awake by some new noise, a recruit coughing, the fire watch changing shifts, the distant sound of aircraft from the nearby airport. Even in these moments of supposed rest, our bodies were being conditioned to maintain a state of constant readiness.

The Korean concept of "jung shin", mental strength and spirit, that my mother had tried to instill in me took on new meaning. This wasn't just about individual fortitude anymore; it was about becoming part of something larger, about subordinating personal comfort and identity to a greater purpose.

Somewhere between consciousness and sleep, I remembered Mr. Redwine's words. The Malibu had needed to be stripped down completely before it could be rebuilt. Now it was my turn, to be broken down, stripped of civilian habits and thinking, ready to be rebuilt into something stronger, something more disciplined, something worthy of the title United States Marine.

Those yellow footprints were more than paint on concrete. They were the first page of a new chapter, the beginning of something that would reshape my body, mind, and entire understanding of who I could become. As sleep finally overtook me, I realized that the person who had stepped onto those footprints would never exist quite the same way again.

Black Friday

The relative calm of receiving week shattered on "Black Friday", the day we met our permanent Drill Instructors. After a week of processing and preliminary training, we sat in perfectly aligned rows on the squad bay floor, sixty recruits arranged by height, awaiting the team that would transform us from civilians into Marines.

Captain Ahern, the company commander, stood before us delivering what felt like a preparatory speech, though his words barely registered through our collective anxiety. The atmosphere was charged with anticipation, like the heavy air before a storm. Then it happened, the duty hut door exploded open with a bang that made every recruit flinch.

Staff Sergeant Lahey emerged like a force of nature. Six feet four inches of perfectly pressed olive drab utility uniform topped by the iconic campaign cover, he moved with the controlled violence of a predator. Most striking was his right hand, the index finger missing, leaving a constant knife-hand gesture that would become the symbol of our next three months. The silence as he surveyed us was absolute, broken only by the ragged breathing of terrified recruits.

"SIT UP STRAIGHT AND LOOK AT ME RIGHT NOW!" His voice filled the squad bay like a physical presence. "MY NAME IS STAFF SERGEANT LAHEY, AND I AM YOUR SENIOR DRILL

INSTRUCTOR. I AM ASSISTED IN MY DUTIES BY STAFF SER-
GEANT MORRIS AND STAFF SERGEANT CHESTNUT."

As if summoned by their names, the other two Drill Instructors appeared. Staff Sergeant Morris, dark-skinned and muscular, moved with
the deadly grace of his Force Recon background. Staff Sergeant Chestnut, though shorter, radiated an intensity that seemed barely contained
by his uniform. They took their positions flanking SSgt Lahey, creating a
trinity of terror that would dominate our lives for the next twelve weeks.

"OUR MISSION IS TO TRAIN EACH ONE OF YOU TO BE-
COME A UNITED STATES MARINE. A MARINE IS CHARAC-
TERIZED AS ONE WHO POSSESSES THE HIGHEST MILITARY
VIRTUES. THEY OBEY ORDERS, RESPECT THEIR SENIORS,
AND STRIVE CONSTANTLY TO BE THE BEST IN EVERY-
THING THEY DO."

The speech continued, each word delivered with precision and purpose. But it was more than words; it was a declaration of intent. Everything we had been, everything we thought we knew about ourselves, was
about to be systematically dismantled and rebuilt according to Marine
Corps standards. The Drill Instructors moved through our ranks like
master craftsmen evaluating raw material, their eyes missing nothing.

THE TRANSFORMATION

Physical training began immediately and relentlessly. Push-ups, mountain
climbers, side straddle hops: exercises meant not just to strengthen our
bodies but to break down individual will and build unit cohesion. "YOU
MOVE TOGETHER, YOU BREATHE TOGETHER, YOU EXIST
AS ONE!" SSgt Morris would shout as we struggled through endless
repetitions. The precision required was absolute. Every movement had
to be synchronized, every count in perfect unison.

Staff Sergeant Chestnut specialized in "IT", Incentive Training, the Marine Corps' euphemism for physical punishment. Five minutes of continuous exercise might not sound long, but under his guidance, it became an eternity. "WHEN THE MIND QUITS, THE BODY FOLLOWS," he would say, pushing us beyond what we thought possible. The concept of individual limits began to dissolve under his relentless pressure.

Even simple tasks became exercises in precision and discipline. Making our racks (beds) evolved into a complex ritual of hospital corners and precise measurements. The sheets had to be stretched so tight that a quarter would bounce off them. Simple actions like eating or using the head (bathroom) had to be done according to exact protocols. "THERE IS A RIGHT WAY, A WRONG WAY, AND THE MARINE CORPS WAY," SSgt Lahey would remind us. "GUESS WHICH ONE YOU'RE GOING TO LEARN?"

The cultural differences that had defined my civilian life began to fade under the relentless pressure for uniformity. Two Korean-Americans served in our platoon, both named Kim. SSgt Lahey, with typical military efficiency, dubbed me "Delta" and the other "Charlie", using the phonetic alphabet to distinguish us. For the first time, my Korean identity became a simple operational necessity rather than a defining characteristic.

Language itself changed. Common words acquired new meanings: deck became floor, bulkhead became wall, head became bathroom. Even time changed, we learned to speak in military hours, where 1:00 PM became thirteen hundred hours. This linguistic overhaul was another tool for breaking down our civilian identities and building Marine Corps mindset.

The process of breaking down extended to our most basic assumptions about ourselves. Every habit, every mannerism, every individual trait was subject to scrutiny and correction. We learned to move as one, think as one, respond as one. Individual achievement meant nothing without unit success.

"YOU ARE NOT SPECIAL," SSgt Lahey would remind us daily. "YOU ARE NOT UNIQUE. YOU ARE MARINE CORPS RE-CRUITS, AND THAT IS ALL YOU ARE UNTIL YOU PROVE YOURSELVES WORTHY OF SOMETHING MORE." His words stripped away the layers of individual identity we'd built over years of civilian life, creating a blank canvas for Marine Corps values.

The psychological pressure was constant and calculated. Every aspect of our environment worked to eliminate individual comfort zones. Personal space became a foreign concept. We changed clothes, showered, and performed bodily functions in full view of others. Privacy, like individuality, was a civilian luxury we could no longer afford.

The Drill Instructors deliberately manipulated the hierarchy of needs. Sleep became a precious commodity, carefully rationed to keep us in a constant state of physical and mental strain. Meals were consumed in precisely timed intervals, each bite taken according to strict protocols. "YOU WILL EAT NOW, YOU WILL SLEEP NOW, YOU WILL BREATHE WHEN I TELL YOU TO BREATHE!" The control extended to every aspect of our existence.

SSgt Morris, during rare moments of relative calm, would explain the psychology behind the training. "We break you down not because we enjoy it, but because we need to rebuild you stronger. A Marine isn't just a trained civilian. We've completely remade him." His Force Recon background gave these words particular weight.

The collective experience began to forge unexpected bonds. Despite the prohibition on personal conversations, we developed silent ways of supporting each other. A subtle nod of encouragement, an extra effort during team tasks, the unspoken understanding that we were all going through this together. These small gestures of solidarity became crucial to our survival.

As the first week under our permanent Drill Instructors drew to a close, we were already changing. The civilians who had stepped onto those yellow footprints were vanishing, layer by layer. In their place, something new was beginning to form: not yet Marines, but no longer purely civilians either.

Combat training began with the building blocks of Marine Corps combat, LINE (Linear Infighting Neural Override Engagement). In the sweltering heat of the San Diego summer, we learned the science of controlled violence. Every movement had purpose, every technique a specific application. Staff Sergeant Morris, drawing on his Force Recon background, brought these lessons to life with a precision that made the abstract concrete.

"A Marine's body is a weapon," he would explain, demonstrating each movement with lethal grace. "But a weapon is only as effective as the discipline behind it." We learned to punch, block, and counter with mechanical precision. The techniques weren't just about fighting, they were about developing the mental discipline that defined Marine Corps combat effectiveness.

Pugil stick training introduced us to controlled combat. Padded helmets, groin protection, and heavy sticks simulating rifles became our first taste of warrior training. The matches were intense, physical proof of who we were becoming.

The rifle range became our new sanctuary. The M16A2 service rifle was introduced to us not just as a weapon, but as an extension of Marine Corps philosophy. "This rifle is built on precision," our Primary Marksmanship Instructors (PMIs) explained. "Every component, every movement, every breath affects your accuracy."

We learned the weapons handling through endless repetition. "Trigger squeeze, breathing control, sight alignment, sight picture," became our mantra. Each component had to work in perfect harmony for optimal performance.

Days on the range started before dawn. We learned to shoot in various positions, prone, sitting, kneeling, standing. Each position required perfect body alignment, controlled breathing, and absolute focus. The PMIs moved among us like technical specialists, adjusting our form with minute corrections that could mean the difference between hit and miss.

"Remember your BRASS," they drilled into us: Breathe, Relax, Aim, Squeeze, Surprise. The fundamentals of marksmanship became another language, another way of thinking. The change was happening not just in our bodies, but in our minds, learning to think like Marines, to approach every task with disciplined precision.

Qualification day brought its own intense pressure. Starting at the 200-yard line in the pre-dawn hours, we worked our way through each distance: 300-yard and then 500-yard line. The targets seemed impossibly small, but the training took over. At the 500-yard line, cover the silhouette with the front sight post. Breath control, trigger squeeze, follow-through. Each shot was a test of everything we'd learned. The sound of firing was punctuated by the crack of rounds finding their marks, each hit a small victory.

Physical training continued relentlessly. Every morning began with PT, pushing our bodies beyond what we thought possible. The runs grew longer, the exercises more demanding. Staff Sergeant Chestnut seemed to take particular pleasure in finding new ways to test our endurance. "Pain is weakness leaving the body!" became more than a slogan; it was our daily reality.

Close Order Drill transformed us from individuals into a unified whole. Hours spent on the parade deck under the merciless sun, learning to move as one. "LEFT! LEFT! LEFT! RIGHT! LEFT!" The cadence became our heartbeat, the movements ingrained at a cellular level. Every step, every turn, every movement had to be precise, synchronized, perfect. It was another form of breaking down individuality to build unit cohesion.

Swimming qualification presented unique challenges. Combat water survival training at the base pool tested not just our physical abilities but our mental fortitude. Full utility uniforms, boots, and rifles added weight and complexity to every movement. Some recruits who had never seen a pool before faced their fears under the watchful eyes of swim instructors who accepted no excuses.

The change was visible in our bodies: lean, hardened, moving with new purpose. But deeper changes were happening inside. Marine Corps discipline was replacing civilian thought patterns. Every task, no matter how small, demanded absolute attention and precision. "Slow is smooth, smooth is fast" became more than a slogan; it was our operating principle.

SSgt Lahey's missing index finger became a symbol of what we were becoming. That perpetual knife-hand, always ready to point out our failures or direct our movements, reminded us that imperfection didn't equal inability. His own adaptation to injury demonstrated the Marine Corps principle of improvising, adapting, and overcoming.

By week six, we were no longer the uncertain recruits who had arrived on those yellow footprints. The physical challenges had reshaped us, but more importantly, they had reshaped our understanding of what we were capable of. Every completed challenge, every qualified test, every mastered skill added another layer to who we were becoming.

In the sixth week of training, our platoon faced a turning point during a particularly grueling force march. The San Diego sun beat down mercilessly as we trudged through the hills of Camp Pendleton, each of us carrying full combat loads. About five miles into the eight-mile march, Recruit Johnson, a recruit from rural Mississippi, began to falter. His steps grew unsteady; his breathing labored beyond normal exertion.

Carrying Each Other

Staff Sergeant Morris noticed immediately. Instead of singling Johnson out for weakness, he barked an order that would become a defining moment for our platoon: "Distribute his load! I said DISTRIBUTE HIS LOAD, Marines!" Without breaking stride, those nearest to Johnson instinctively reached for his gear. I found myself taking his rifle, another recruit shouldered his pack. Within seconds, Johnson's burden was shared among six of us.

"This is what Marines do," SSgt Morris called out, his voice carrying over our rhythmic footfalls. "We carry each other. No Marine falls behind because no Marine stands alone." The words sank deep, resonating with something fundamental that was taking shape within us. We finished that march not as individuals struggling separately, but as a unit moving with single purpose.

Our platoon's evolution from a collection of individuals into a cohesive unit happened gradually, almost imperceptibly. Like the careful assembly of an engine, each component had to be precisely fitted, each connection strengthened through shared experience. The cultural barriers that had seemed so significant in civilian life, race, background, education, dissolved under the greater pressure of collective survival.

During evening cleanup duty, conversations would emerge in whispered tones as we worked. Hernandez, a former gang member from East Los Angeles, shared stories of street life while meticulously polishing his boots. Williams, a college dropout from Boston, recited Shakespeare as we scrubbed toilets. My own experience as a Korean-American mechanic and pizza guy from Riverside became just another story in our platoon's mix.

The Drill Instructors understood this bonding process intimately. They would create challenges that could only be overcome through teamwork. During one memorable afternoon, SSgt Chestnut introduced us to the "incentive squad bay." The entire platoon had to complete a complex series of exercises, moving as one unit. If a single recruit fell out of sync, we all started over.

"You're only as strong as your weakest link," SSgt Chestnut bellowed as we struggled through endless repetitions. "But a chain that knows how to support its links becomes unbreakable." Hour after hour, we learned to anticipate each other's movements, to sense when someone was struggling and compensate automatically. The physical exhaustion broke down our individual barriers, forcing us to rely on each other in ways we'd never experienced.

Language barriers, which had once seemed insurmountable, became opportunities for connection. Rodriguez, who spoke limited English, excelled at weapons maintenance. He taught through demonstration, his hands moving with practiced precision as he field-stripped his rifle. In return, others helped him with English pronunciation during mail call. I found myself translating not just words but concepts, drawing on my experience of moving between cultures.

Even our failures brought us closer. When Recruit Cooper struggled with his rifle qualification, the entire platoon stayed after hours to help him prepare. We took turns coaching him through sight alignment, breathing techniques, trigger control, each of us contributing what we'd learned. When he finally qualified, his success felt like a victory for us all. The Drill Instructors watched this evolution with knowing eyes, understanding that breaking us down individually was only the first step. The real goal was building us up collectively, creating a unit that was stronger than the sum of its parts.

Night fire training became another test of unity. In the darkness of the range, we had to rely completely on each other. Safety depended on absolute trust, on knowing that every Marine around you understood and followed the same protocols. The bright traces of rounds heading downrange painted streaks against the night sky, each shot proof of our collective discipline.

The deepest moments of unity often came during the quietest times. During firewatch in the dead of night, two recruits would patrol the squad bay in two-hour shifts. These midnight hours became sacred spaces of connection. Whispered conversations in the darkness revealed dreams, fears, and determination. We learned each other's stories, not as civilians had told them, but as Marines were beginning to shape them.

Staff Sergeant Lahey observed our growing cohesion with subtle approval. "A platoon becomes a family," he told us during a rare quiet moment. "Not because you like each other, though you might. Not because you're similar, because you're not. But because you've learned to put the unit above yourselves. That's what makes Marines different."

EARNING THE TITLE

As the tenth week of training approached, a palpable shift occurred in our platoon. The end was in sight, but the final hurdles would test everything we'd learned about ourselves and each other. The Drill Instructors' intensity, rather than diminishing, reached new heights. Staff Sergeant Lahey gathered us in the squad bay one evening, his missing index finger emphasizing each point as he spoke.

"The next two weeks will define you," he declared, his voice carrying its usual authority but with an underlying current of gravity we'd learned to recognize. "Everything you've learned, everything you've become, it all culminates now. Some of you will discover you're stronger than you ever imagined. Others will find out if you truly belong here."

The final Physical Fitness Test (PFT) loomed before us like a mountain to be climbed. Three simple but demanding events: pull-ups, crunches, and a three-mile run. We'd been training for this throughout boot camp, but now it carried the weight of qualification. The morning of the test arrived with an unusual silence in the squad bay, each recruit focused inward, mentally preparing for the challenge ahead.

The pull-up bar became our first battleground. One by one, we approached it, each Marine drawing on reserves of strength built over weeks of training. The counting was precise, each repetition scrutinized by watchful eyes. Each pull-up had to be perfect, full extension to full contraction. I achieved 20 pullups, 100 points.

The sit-ups followed, two minutes of continuous motion that tested not just abdominal strength but mental fortitude. Partners held feet, counted repetitions, encouraged through gritted teeth. The sound of bodies moving in unison against the mat created a rhythm, a cadence of determination. Some recruits pushed through pain, others fought fatigue, all drove forward toward the minimum requirement and beyond. I hit 80 sit-ups, also 100 points.

The three-mile run was our final challenge. Starting in the pre-dawn hours, we lined up on the track. The San Diego marine layer hung low, creating an ethereal atmosphere as we began. The run wasn't just about speed; it was about pacing, about understanding your body's limits and pushing past them. As we circled the track, the fog began to lift, both literally and metaphorically.

Each lap brought us closer not just to the finish line, but to our goal of becoming Marines. I ran a 18:10, 99 points, bringing my total to 1 point short of a perfect 300 score. That was the highest I ever scored on a PFT in my entire time in service, my very first one that counted.

The close order drill evaluation tested our unity one last time. Hours spent on the parade deck came down to a single performance. Every step, every movement had to be precise and synchronized. Staff Sergeant Morris watched with an expert's eye as we executed the manual of arms. The sound of rifles moving in perfect unison echoed across the parade deck - sixty recruits moving as one body, one mind.

Then came the moment we'd all been anticipating: the arrival of our families for the Family Day ceremony. After eleven weeks of isolation from the civilian world, seeing familiar faces triggered an unexpected wave of emotions. Parents, siblings, spouses, they saw not the people who had left home nearly three months ago, but transformed beings who stood straighter, moved with purpose, carried themselves with newfound dignity.

My mother's face registered shock when she first saw me. The physical change was dramatic: thirty pounds lighter, all trace of civilian softness replaced by lean muscle. But it was more than that. She later told me she saw in my eyes something she'd never seen before: an unshakable confidence, a clarity of purpose that transcended our cultural expectations.

Graduation Day

The graduation ceremony itself was a blur of precision and pride. As we stood on the parade deck, the morning sun glinting off our brass insignia, Staff Sergeant Lahey's voice carried across the formation one last time: "Platoon 1070, you are no longer recruits. You have earned the title United States Marine. Never forget the price of that title, or the responsibility it carries."

The moment we had trained for arrived with three simple words: "Platoon, ten-hut!" We snapped to attention as one. "Platoon 1070, dismissed!" In perfect unison, we took one step back, executed a precise about-face, and for the first time in twelve weeks, broke ranks as United States Marines.

Later, as families mingled on the parade deck, I caught sight of our Drill Instructor team standing apart, watching their latest platoon complete its cycle. Staff Sergeant Lahey's missing index finger rested against the brim of his campaign cover in an understated salute. Staff Sergeant Morris stood with his characteristic quiet intensity, while Staff Sergeant Chestnut allowed himself a rare smile.

My parents approached. My mother reached up to touch my cheek, as if verifying that I was real. "우리 아들," she whispered, my son, but the words carried new weight. My stepfather stood awkwardly to the side, our previous conflicts somehow smaller now.

Across the parade deck, Hernandez's former gang members stood in respectful silence. Williams' college professor father beamed with unexpected pride. Each reunion was its own story.

Behind us, a new bus was pulling in. I could see them through the window, forty young men about to step onto those yellow footprints for the first time. They looked exactly like we had twelve weeks ago. Uncertain. Afraid. Ready to be broken down and rebuilt.

I didn't watch them get off the bus. I had somewhere else to be.

CHAPTER 3

BECOMING A MARINE

061200ZNOV92

The title "Marine" came with graduation but becoming one would take longer. The thirteen weeks between boot camp and my first duty station were a strange limbo. I was no longer a recruit but not yet part of the fleet. Thirty days of leave sent me back to Riverside wearing a uniform that fit differently than the civilian clothes I'd left behind. Then came the Hometown Recruiting Assistance Program, where I walked the halls of my old high school as proof that troubled kids could become something more. Marine Combat Training at Camp Pendleton taught me that every Marine, regardless of job, must be a rifleman first. And Supply School at Camp Johnson, North Carolina, revealed something unexpected: that the logistics of war, the tracking of bullets and beans and bandages, would become my specialty and my purpose. These were the weeks that transformed a boot camp graduate into something resembling an actual Marine.

THIRTY DAYS OF PERSPECTIVE

After graduation from boot camp, the Marine Corps granted me thirty days of leave before my next phase of training. This period would include ten days of recruiting duty, but first, I had nearly three weeks to process everything that had happened in just thirteen weeks at MCRD San Diego.

The drive back to Riverside felt surreal. The familiar streets looked somehow smaller, as if the perspective gained through Marine Corps training had literally changed how I viewed my old world. Even sitting in the driver's seat felt different, my posture automatically straighter, my movements more deliberate, my awareness of surroundings more acute.

Each morning, I rose before dawn, maintaining the discipline instilled at boot camp. Running through the quiet streets of Riverside, past the high school, past the pizza shop, past all the familiar landmarks of my previous life, I felt how much I had changed. Other teenagers were still sleeping, but I was a Marine now, and Marines PT every morning, whether required to or not.

The dynamic with my parents had shifted noticeably. My mother, who had fought so hard against my enlistment, now spoke proudly to her friends at church about her Marine son. My stepfather's usual authoritarian stance had softened into something approaching respect. The uniform seemed to bridge cultural gaps that had once seemed insurmountable. Even our conversations felt different, less confrontational, more measured, as if my military service had finally proven something they had been waiting to see.

Meetings with high school friends proved enlightening and sometimes disheartening. Many seemed stuck in the same patterns: working the same jobs, hanging out at the same places, talking about the same dreams without taking steps to achieve them. I'd learned that over 80% of people live and die within 40 miles of where they grew up. Now I saw that statistic play out before my eyes. Friends who had once mocked my decision to join the Marines now looked at me differently, some with respect, others with what seemed like envy.

Meeting Jenny, Again

It was during an afternoon at Tyler Mall when the past and present collided unexpectedly. I was in my Dress Deltas, the uniform still crisp and new, when I heard a familiar voice call my name. Jenny stood there, looking much as she had in high school, beautiful, graceful, with that same long black hair, but now she seemed somehow smaller, less significant than the larger-than-life presence she had held in my teenage heart.

"David...you look amazing," she said, her eyes taking in the uniform. The conversation that followed was awkward, full of unspoken words. She suggested getting lunch, catching up, perhaps rekindling what we had lost. Her eyes held a mixture of admiration and regret, so different from the last time I'd seen her, when she had chosen my friend over me.

But I was no longer the heartbroken teenager who had partly joined the Marines to escape the pain she had caused. The uniform represented how far I had come, choices made and paths taken. With a politeness that surprised even me, I declined her invitation.

"The past should stay in the past," I told her, the words coming easily, without bitterness or regret. The Marine Corps had taught me about moving forward, about the difference between holding onto memories and being held back by them. Walking away from that encounter, I felt a final piece of my civilian past settle into place, not forgotten but properly filed away in the archive of experiences that had shaped me.

These encounters with family, old friends, and past relationships showed how much I had changed. The Marine Corps had given me something beyond physical conditioning and military bearing: a clarity of purpose, a sense of direction that stood in stark contrast to the uncertainty I saw in many of my peers.

Even my family's restaurant felt different. Where once I had seen it as a symbol of immigrant constraints, I now recognized it as part of the foundation that had made me strong enough for military service. The discipline required to run a small business, the attention to detail needed in food preparation, the importance of customer service, all had contributed to my success in the Marines.

As the first portion of my leave drew to a close, I prepared for the next phase, the Hometown Recruiting Assistance Program (HRAP). The perspective gained during these weeks proved valuable as I returned to my old high school, not as a former student, but as a Marine with a mission to help others see the possibilities that service could offer.

RIVERSIDE RECRUITING

GySgt Reefer had arranged for me to participate in the Hometown Recruiting Assistance Program (HRAP). This voluntary duty allowed new Marines to return home briefly before their next phase of training, helping recruiters connect with potential candidates. For me, it meant returning to Riverside Poly High School, not as the uncertain student I'd been just months ago, but as a United States Marine.

Mr. Redwine's Handshake

My first stop was Mr. Redwine's auto shop. The familiar smell of motor oil and metal brought an unexpected wave of nostalgia. Through the open bay door, I could see him bent over an engine, his massive frame instantly recognizable. He straightened up as I approached, wiping his hands on his signature red shop rag. His eyes took in my uniform, lingering on the Eagle, Globe, and Anchor emblem that I'd fought so hard to earn.

"Well, look who's back," he said, his voice carrying a warmth that transcended the formal greeting. Then he did something unexpected, he extended his hand. Not the casual handshake of a teacher greeting a former student, but the firm grip of one military member acknowledging another. "Welcome home, Marine."

Our conversation flowed naturally, bridging the gap between my past and present. Mr. Redwine shared stories from his own Navy experience that he'd never mentioned before, as if my new status had opened a door to deeper understanding. We talked about Vietnam, about the Marines he had served with on ship, about how the Navy changes you in ways civilians can never fully understand.

"The discipline you showed rebuilding that Malibu," he said, gesturing toward the parking lot where my car sat, "that's what made me sure you'd make it through boot camp. A Marine isn't made in twelve weeks - he's made through years of building the right habits, the right mindset. You were becoming a Marine long before you stepped on those yellow footprints."

The contrast with other teachers' reactions was stark. Walking the halls with GySgt Reefer, I encountered Mr. Ayres, my former history teacher. His response crystallized the civilian-military divide: "I'm disappointed," he said, his voice heavy with judgment, "that you've chosen killing babies over getting an education."

The comment struck me like a physical blow, but my reaction surprised even me. Where once I might have shrunk from such confrontation, my Marine Corps training kicked in. I maintained my bearing, explaining calmly and professionally about the educational benefits, the technical training, the opportunities for personal growth. His dismissive response said more about his prejudices than my choices.

The day continued with classroom visits, each one a study in contrasting reactions. Some teachers welcomed me warmly, genuinely interested in what I'd become. Others barely concealed their disapproval, seeing the uniform as a symbol of something they opposed rather than the mark of personal achievement it represented.

But it was the students' reactions that proved most enlightening. Their questions went beyond the usual inquiries about boot camp difficulty or combat expectations. They wanted to know how someone they'd known as a quiet Korean-American kid had become this confident Marine standing before them. My responses drew from fresh experience: the challenges, the growth, the discovery of strength they didn't know they possessed.

In the pizza and ribs restaurant, my parents' reaction to my uniformed presence was complex. Pride warred with lingering uncertainty about my choice. The customers, however, responded differently. Regular patrons who'd known me as the owner's quiet son now saw something else, a Marine. The respect in their interactions was palpable, though sometimes awkward, as if they weren't quite sure how to bridge the gap between their memory of who I was and who I'd become.

GySgt Reefer observed these interactions with knowing eyes. "This is why we do HRAP," he explained during a quiet moment. "It's not just about recruiting; it's about showing what's possible. You're living proof of what the Marine Corps can do for someone who's willing to commit to the process."

The ten days passed quickly, each interaction reinforcing how much I'd changed. The uniform I wore wasn't just a change of clothes; it was a symbol of everything I'd become, everything I was still becoming. Mr. Redwine understood this better than anyone. Our final conversation before I left for Marine Combat Training resonated with deeper meaning.

"Remember," he said, once again extending that military handshake, "the hardest part isn't becoming a Marine, it's living up to what that title means every single day. Keep the same attention to detail you showed in this shop. It'll serve you well."

Those first days after boot camp marked a unique period of limbo in Marine Corps life. No longer recruits but not yet fully integrated into the Fleet Marine Force, we existed in a transitional space between our civilian past and military future. The ten days of recruiting duty had given me a taste of my new identity, but returning to the military environment brought fresh challenges.

EVERY MARINE A RIFLEMAN

Camp Pendleton's 52 Area emerged from the morning fog like a different world from the recruit training facility we'd left behind. The mountains loomed in the background, their shadows stretching across training grounds that would become both classroom and proving field for the Marine Corps' fundamental truth: every Marine, regardless of eventual specialty, is first and foremost a rifleman.

The contrast with boot camp was immediately apparent. Where recruit training had focused on breaking down civilian identity and instilling basic discipline, MCT assumed that foundation and built upon it. Our instructors, while maintaining high standards, spoke to us as Marines rather than recruits. This subtle shift reflected our new status and the elevated expectations that came with it.

"Listen up, Marines," our lead instructor, Corporal Ennis, announced during our first formation. "What you learned in boot camp was how to become Marines. Here, you'll learn how to be Marines in combat. Every one of you, whether you're going to be a cook, a computer specialist, or a supply clerk, needs to master these skills. Because when the rounds start flying, your MOS doesn't matter, your ability to fight does."

The training schedule reflected this combat-centric philosophy. Our days began with combat conditioning, runs in full gear, tactical movements through Pendleton's harsh terrain, and physical training that emphasized combat functionality over mere fitness. The weight of our flak jackets, Kevlar helmets, and combat loads became constant companions, transforming every movement into an exercise in endurance.

The M16A2 rifle took on new meaning here. In boot camp, we'd learned marksmanship as a technical skill. At MCT, we learned to treat our weapons as extensions of ourselves. Combat shooting introduced new challenges, firing from unstable positions, engaging moving targets, maintaining accuracy under physical stress. The precision I'd developed in boot camp had to adapt to these dynamic conditions.

"Your rifle," our combat instructor explained while demonstrating tactical reloads, "is like any other tool. Understanding its basic operation is just the start. You need to master it under any condition, in any situation."

Tactical training introduced us to the complexity of modern warfare. We learned fire team movements, squad tactics, and the fundamentals of urban combat. The mechanical precision required in these exercises went far beyond anything in boot camp. Each movement had to be coordinated, each position covered, each action executed with both speed and control.

Night operations added another dimension to our training. Moving through Pendleton's terrain in darkness, using nothing but moonlight and the occasional flash of illumination rounds, we learned to trust our training and each other. The darkness stripped away individual identity, forcing us to rely on the uniformity of our training and the solidarity of our unit.

Communication became critical in ways we hadn't experienced before. Hand signals, radio protocols, and tactical reports had to be mastered. The precision of language took on life-or-death importance, every word, every gesture carrying meaning that could impact an entire operation. This wasn't just about following orders anymore; it was about participating in a complex tactical dialogue.

The culminating field exercise brought everything together. For three days, we operated in a simulated combat environment, conducting patrols, responding to contact, and maintaining security in an ever-changing tactical situation. Sleep was minimal, conditions were harsh, and the pressure was constant. This wasn't about breaking us down, as boot camp had been; it was about proving we could function as Marines under combat conditions.

Throughout the training, the systematic thinking I'd developed translated into excellent weapons handling and tactical awareness.

As MCT drew to a close, we understood more deeply what it meant to be Marines. The title we'd earned in boot camp now carried additional weight, the weight of combat readiness, of tactical responsibility, of being prepared to fight regardless of our eventual specialties. We were no longer just Marines in name; we were Marines prepared for the fundamental mission of our Corps.

The final words from our combat instructor stayed with us: "Remember, Marines, your primary MOS is 0311 (rifleman). Everything else is secondary. Carry that knowledge with you, maintain these skills, because you never know when you'll need them."

THE BACKBONE OF WAR

Camp Johnson, North Carolina presented a stark contrast to the combat-focused environment of MCT. Located near Jacksonville, the base's Supply School represented a return to technical precision, but with a military twist that transformed seemingly mundane administrative tasks into critical components of Marine Corps readiness.

The five-week course introduced us to an axiom as old as warfare itself: "Amateurs talk tactics, professionals talk logistics." Our instructors, seasoned supply Marines, emphasized this point repeatedly. "You can have the best-trained combat units in the world," our lead instructor explained during our first class, "but without proper supply support, they're just well-trained individuals standing still."

The connection between mechanical systems and supply chains became immediately apparent. Just as an engine requires precisely timed fuel delivery and properly flowing oil, military units require carefully managed supply lines. I'd learned to see systems as interconnected networks; now I was applying that understanding to military logistics.

SASSY (Supported Activities Supply System) became our new language. This mainframe-based supply management system was the nerve center of Marine Corps logistics. Learning its intricacies required absolute precision. Each transaction code, each supply category, each procedural step had to be mastered perfectly.

"Think of SASSY as an engine," our systems instructor suggested. "Each component, requisitioning, inventory management, distribution, has to work in perfect harmony. One mistake in the system can cause the entire supply chain to seize up, just like a thrown rod can destroy an engine."

The curriculum was intensely technical. The 4400-123 SASSY Manual became our bible. We learned supply accounting, inventory management, requisition procedures, and the complex web of regulations governing military property. Every morning began with hands-on computer training, learning to navigate the green-screen terminals that would become our primary tools. The afternoon brought practical exercises in warehouse operations, teaching us the physical aspects of supply management.

My technical aptitude paid off. Where others struggled with the systematic nature of supply procedures, I found familiar patterns. The precision required for proper supply documentation matched what I'd learned about attention to detail. Inventory management demanded careful measurement and exactness.

The shift from combat training to supply school revealed another aspect of Marine Corps versatility. We were still Marines, the physical training continued, the military bearing remained sharp, but now we were developing specialized skills that would support the Corps' broader mission. Our instructors emphasized that supply Marines, while perhaps not on the front lines, were crucial to mission success.

"Every bullet fired, every meal eaten, every vehicle moved, it all depends on supply," our warehouse operations instructor stressed. "You're not just pushing paper or moving boxes. You're ensuring that Marines in combat have exactly what they need, when they need it."

The course culminated in a practical exercise that tested both our technical knowledge and our ability to perform under pressure. Given a simulated combat scenario, we had to establish and maintain a supply chain while dealing with changing priorities, equipment failures, and the fog of war. The exercise demonstrated how our seemingly administrative skills could directly impact combat effectiveness.

The final weeks of Supply School delved deeper into the intricate technical systems that powered Marine Corps logistics. Every component had its purpose, and precision was paramount.

The SASSY terminal became my new workbench. Instead of wrenches and timing lights, I now wielded transaction codes and inventory reports. The green phosphor display might have seemed primitive compared to modern computers, but it represented a sophisticated system that tracked millions of items across the global Marine Corps supply chain. A single mistyped character along the 80-character columns could disrupt an entire supply request.

"The mainframe doesn't forgive mistakes," our systems instructor emphasized, walking us through complex requisition procedures. "You can't estimate supply codes. Precision isn't just about being accurate; it's about maintaining the integrity of the entire system."

The inventory management system revealed itself as a masterpiece of organizational logic. We learned to track everything from individual rifle parts to entire vehicles, each item categorized with specific supply codes that determined how it would be requisitioned, stored, and distributed. Multiple components working in precise synchronization to achieve a larger purpose.

Our training in supply forecasting made immediate sense to me. We learned to anticipate supply requirements based on operational tempo and historical data. Every unit's supply demands created patterns that, once understood, could be used to prevent shortages before they occurred.

"Think of it like preventive maintenance," explained our forecasting instructor. "You don't wait for units to run out of crucial supplies before processing requisitions. You anticipate and prepare."

The technical documentation requirements were exhaustive. Every transaction required multiple forms, each with its own specific format and procedure. The military's obsession with proper documentation, which had seemed excessive at first, began to make sense. In an organization as vast as the Marine Corps, standardized documentation wasn't just bureaucracy; it was the only way to ensure smooth operations across multiple units and locations.

We learned to handle complex supply codes that formed their own specialized language. National Stock Numbers (NSNs), Federal Supply Classifications (FSCs), and Unit of Issue codes had to be memorized and applied with absolute precision. Each thirteen-digit NSN told a story about an item's origin, purpose, and place in the supply chain. It was like learning to read the DNA of military logistics.

Physical warehouse operations integrated seamlessly with our technical training. We learned that efficient storage and retrieval systems were as crucial as accurate computer records. The warehouse became another type of operations center, with inventory flowing through a carefully managed system. Everything had its place, and efficiency depended on maintaining precise order.

"Your technical skills are your tools," our lead instructor said during our final week. "But your understanding of the system, how everything works together, how to adapt when things go wrong, that's what makes you a Supply Marine."

The final days at Camp Johnson carried the weight of anticipation. As we completed our last technical assessments and prepared for graduation, the topic on everyone's mind was orders, those official documents that would determine our next duty stations and, in many ways, the trajectory of our Marine Corps careers.

The morning my orders arrived felt like opening a door to the future. Standing in the company office, I received the manila envelope with "LCPL KIM, DAVID 7824" stamped across the front, my Rank, Last Name, First Name, and last four of my SSN. My heart raced as I carefully opened it, the crisp sound of tearing paper echoing in the quiet room. The words jumped off the page: "CAMP PENDLETON, 1ST SUPPLY BATTALION, SUPPLY COMPANY, SASSY MANAGEMENT UNIT (SMU)."

Camp Pendleton. The SASSY Management Unit wasn't just any supply assignment; it was the nerve center of West Coast Marine Corps logistics operations. Only the top graduates of the course received orders to the SMU. In this course, I was one of only three, the other two went to Camp Lejeune and Okinawa.

Our graduation ceremony was markedly different from boot camp. The small parade deck at Camp Johnson lacked the grandeur of MCRD San Diego, but the commanding officer's words carried weight: "You are now technical experts entrusted with maintaining the systems that keep our Corps functioning. Every supply request you process, every inventory you manage, directly impacts Marines in the field."

CHAPTER 4

FIRST DUTY STATION

280800ZMAR93

Camp Pendleton sprawled across 125,000 acres of Southern California coastline, a world unto itself where I would spend the next three years learning what it actually meant to be a Marine. Boot camp had forged the identity. Supply School had given me a specialty. But the fleet was where theory met reality, where a lance corporal with a supply MOS would either prove himself or fade into the background. I arrived in March 1993, eighteen years old, eager to demonstrate that my test scores and training meant something. What I found was a technical kingdom called the SASSY Management Unit, a commanding officer who saw potential I hadn't yet recognized in myself, a barracks culture that ran on cheap beer and weekend liberty, and a Korean church in Riverside where I would meet the woman who would become my first wife. The next three years would transform me from a boot Marine into a husband, a technical innovator, and eventually, a combat-tested NCO.

ARRIVAL

Camp Pendleton emerged from the California coastal landscape like a world unto itself, 125,000 acres of military infrastructure that would become my entire universe. Driving through the main gate in my Malibu, Service Alpha uniform pressed to perfection, I felt the weight of transition. This wasn't the training environment I'd experienced during MCT; this was the real Marine Corps, where I would put all my training into practice.

The 22 Area sprawled before me, a complex of barracks buildings and administrative offices that housed Supply Company and its various units. The March morning fog still clung to the hills, creating an almost mystical atmosphere as Marines moved with purpose between buildings. The sound of morning physical training echoed from somewhere nearby, the familiar cadence of Marines running in formation providing an oddly comforting soundtrack to my arrival.

The Supply Company office embodied military precision. Inside, the administrative clerk, a Lance Corporal with a nametag reading "MARTINEZ", processed my check-in paperwork with mechanical efficiency. Each form required multiple signatures, each step following a precise protocol that reflected the attention to detail I'd come to expect from Marine Corps administration.

"SASSY Management Unit?" Martinez looked up from my orders, a flicker of recognition crossing his face. "That's Captain Murney's domain. You'll want to make a good first impression there, Marine. She's a hardass." His emphasis on the captain's name carried a weight I wouldn't fully understand until later.

The walk to my assigned barracks was a study in military geography. Each building, each walkway, each gathering space had its purpose. The barracks itself, building 22103, was a three-story concrete structure that had housed generations of Marines before me. The room assignment put me with two other Marines in a modified two-man room, a common arrangement that reflected the constant balance between military efficiency and limited resources.

My roommates were already established: Lance Corporals James Wilson and Frank Massa, both supply warehouse specialists who had arrived a few months earlier. The room reflected their personalities. Wilson's area was organized with almost obsessive precision, while Massa maintained a more relaxed (though still within regulations) approach with Puerto Rican flair.

"You're a '43? SMU?" Massa asked as I stored my gear in my assigned wall locker. "Get ready for some serious computer time, man. That's some crazy whizbang shit there." His casual demeanor carried an undertone of respect for the technical demands of the unit, which worked alongside the 3051 MOS, known as the box kickers.

The SASSY Management Unit occupied a separate building, a sprawling single-story structure that hummed with the sound of computer systems and air conditioning units working to keep the sensitive equipment cool. Walking in for the first time, I was struck by the contrast between the building's plain exterior and the sophisticated technology within.

Captain Murney

Captain Kathleen Murney emerged as the driving force behind the unit's evolution. Standing six-foot-two with flaming red hair and shoulders that suggested a background in competitive athletics, she commanded attention through sheer presence. Her leadership style combined technical expertise with strategic vision, she didn't just want the unit to function; she wanted it to innovate.

Captain Murney's presence was felt before she was seen. The atmosphere in the unit shifted subtly as she moved through the spaces, her tall frame and commanding presence drawing instinctive responses from

Marines at their workstations. When she stopped at my position, I snapped to attention with parade ground precision.

"At ease, Lance Corporal Kim," she said, her voice carrying both authority and analytical sharpness. "Your test scores from supply school were impressive. Let's see if you can apply that knowledge in a real-world environment." Her evaluation was direct, professional, and carried clear expectations of excellence.

My assigned section, Initial Issue Provisioning (IIP), introduced me to the practical application of everything I'd learned in supply school. The mainframe terminals lined up in rows, their green screens displaying the complex coding of supply requisitions and inventory management. It was here that I would begin to understand the true complexity of military logistics.

Sergeant Sweetsir, my direct supervisor, approached technical training with patience. "The systems are complex," he explained, walking me through my first requisition entries, "but they follow logical patterns. Once you understand how the components work together, you can diagnose and solve almost any problem."

The first week established routines that would define my early months at Camp Pendleton. Physical training at 0600, morning formation at 0800, then days filled with learning the intricacies of supply management. Each transaction code, each inventory procedure, each system interface required precision and attention to detail.

Living in the barracks brought its own education. The social dynamics of permanent party Marines differed significantly from the training environment. Relationships were more relaxed but also more complex, with rank and position creating subtle hierarchies that influenced everything from dining facility interactions to weekend liberty plans.

The base itself became a character in my daily life. The coastal location meant mornings often started with marine layer fog that burned off to reveal stunning ocean views. The hills that had challenged us during MCT now served as a constant backdrop to daily activities. The sound of aircraft from the nearby airfield, the distant thump of artillery from the training ranges, the constant movement of Marines and vehicles: all of it became the soundtrack of military life.

THE SASSY MANAGEMENT UNIT (SMU)

The SASSY Management Unit (SMU) operated like a complex nervous system for the Marine Corps' West Coast operations. Hidden behind the unassuming exterior of Building 22145, rows of terminals engaged with the IBM mainframe computers in Albany, Georgia processed an endless stream of supply transactions that kept the entire Pacific region functioning. The constant hum of cooling systems and the soft click of keyboards created a backdrop to what was essentially a technological command center.

My position in Initial Issue Provisioning (IIP) introduced me to the true scope of military logistics. Every new unit being formed, every existing unit being restructured, required a complete analysis of supply requirements. This wasn't just about ordering equipment; it was about understanding the intricate web of dependencies that kept military units operational.

First Lieutenant Donald Nish, our Officer in Charge of IIP and PWR (Prepositioned War Reserve), approached his role with a detachment that initially puzzled me. A soft-spoken Mormon who seemed to be marking time until his military obligation ended, he represented a different kind of military leadership. His administrative focus meant that technical innovation was left largely to the enlisted Marines who understood the systems intimately.

The technological infrastructure itself was a marvel of early 1990s computing. The mainframe systems, with their green phosphor displays and complex command structures, required a deep understanding of both supply protocols and computer operations. Each terminal became a window into the vast network of military logistics, where a single keystroke could affect supply chains across the Pacific theater.

Within three months, my role began to shift. Captain Murney had noticed my ability to grasp both the technical and practical aspects of supply operations. When a special projects position opened up, she moved me from IIP into a role that would allow for greater innovation. "Your technical aptitude is wasted on routine transactions," she explained. "I need someone who can see the bigger picture."

The special projects assignment became my proving ground. The challenge was significant: develop an offline supply tracking system that could operate in environments with limited or no satellite communication. This wasn't just a technical problem; it was a strategic necessity for forward-deployed units that couldn't maintain constant connectivity.

The solution drew on everything I had learned. Working with a small team, we developed a system that could batch process transactions, storing them until communication was restored. The approach required understanding how each component affected the whole, finding ways to maintain functionality under less-than-ideal conditions.

Our testing environment occupied a corner of the SMU, filled with hardware configurations that mimicked field conditions. Hours stretched into days as we refined the system, each improvement bringing us closer to a solution that could work in real-world deployments.

The human element of the work proved as complex as the technical challenges. Each improvement we developed had to be documented and taught to others. I found myself breaking down complex systems into understandable components, using practical examples to illustrate technical concepts.

Captain Murney's support was crucial. She provided resources, defended our innovative approaches to higher command, and most importantly, gave us the freedom to fail and learn from our mistakes. Her leadership style became a model for how technical units could foster innovation while maintaining military discipline.

The impact of our work extended beyond the SMU. Other units began requesting our offline processing system, recognizing its potential for field operations. Each implementation brought new challenges and refinements, as different operational environments revealed both strengths and weaknesses in our approach.

The technical achievements of the unit became a source of pride, but they also highlighted a fundamental truth about military technology: the most elegant solutions come from understanding both the technical capabilities and the practical needs of Marines in the field. Every system we developed, every problem we solved, was ultimately about supporting Marines who depended on reliable supply chains.

LIBERTY

The 22 area Bachelor Enlisted Quarters (BEQ), "barracks", operated under its own unique social ecosystem. Despite the rigid military structure that governed our professional lives, the barracks culture fostered a kind of controlled chaos that defined our off-duty hours. The daily rhythm followed a precise pattern that would have seemed impossible in any other profession.

Physical Training began at 0600 sharp each weekday morning. By 0830, we were at our duty places, the previous night's activities pushed aside in favor of military precision. Lunch breaks became strategic operations, a quick meal followed by what we called "tactical naps" in the warehouse, a skill I quickly mastered. The afternoon shift would dissolve into another nap, and then the real social life would begin.

By 2200, we'd be out in Oceanside or hitting base establishments, seeking the kind of release that only young Marines far from home could understand. Lance Corporal Austin Grenier became my closest companion in this world of structured chaos. Everything about him embodied the stereotype of a young Marine: loud, fearless, with an almost magnetic ability to find trouble and transform it into adventure. Where I approached life with calculated precision, Grenier lived with a reckless enthusiasm that was both terrifying and irresistible.

The 22 area enlisted club and the Del Mar e-club became our regular haunts, a privilege of military life that seemed surreal to civilians. Before the legal drinking age of twenty-one, we had access to establishments where the normal rules appeared not to apply. Each night out became a carefully orchestrated mission, finding female companionship, pushing boundaries, testing the limits of what was acceptable while staying just within the bounds of military discipline.

Captain Murney maintained an acute awareness of these social dynamics. Her counseling about my friendship with Grenier revealed a deeper understanding of young Marine behavior. She recognized the potential pitfalls while understanding that these social bonds were an essential part of military life. Looking back, I suspect her decision to move me from IIP was partially motivated by a desire to separate me from Grenier's influence, though she never explicitly said so.

The partying was more than youthful rebellion; it served as a pressure release valve for young men and women living under extreme discipline. We worked harder than most civilians could imagine, maintaining peak physical and mental performance despite our nighttime activities. The ability to drink until 0400, catch a few hours of sleep, then function at full capacity became a point of pride, though it wasn't sustainable in the long term.

Our weekend rituals followed a predictable pattern. Friday nights usually started at the enlisted clubs, where the drinks were cheap and the music was loud. The crowd was a mix of Marines from different units, each group carrying its own reputation. The Del Mar e-club was particularly known for its high ratio of single women. The supply Marines were known for their technical precision during the day and their ability to party at night, a contradiction that somehow made perfect sense in the military environment.

As the night progressed, we'd often move to establishments in Oceanside. The town existed in a symbiotic relationship with the base, its businesses catering to Marine paydays and youthful enthusiasm. We learned which places to avoid, which bars were friendly to Marines, and most importantly, how to handle the complex social dynamics between civilians and military personnel. Oceanside even hosted an 18+ adult entertainment club that was obviously catering to those young Marines.

Yet beneath the surface of this seemingly endless party, I felt an emptiness that no amount of social activity could fill. As 1993 drew to and end, while the technical challenges of the SMU provided intellectual stimulation, the constant cycle of work and partying felt increasingly hollow.

The contrast between my professional and social lives grew starker with each passing month. During the day, I was developing sophisticated supply systems and earning recognition for technical innovation. At night, I was part of a culture that seemed stuck in a loop of temporary pleasures and superficial connections. This dichotomy would eventually lead to changes in my social circles, but not before teaching valuable lessons about balance and personal growth.

KATHY

The dating scene for a young Marine in the early 1990s presented unique challenges, especially in matters of the heart. My first significant encounter, outside of those casual flings that lasted only a day or two, came during a Fourth of July celebration in 1993. Kimberly was everything that embodied youth and beauty in Southern California - nineteen years old, five-foot-seven, with the kind of curated perfection that came from years of pageant training. Initially, she was dating my friend Austin Grenier, but when that relationship ended, her attention shifted to me with a directness that was both flattering and unsettling.

"I've never been with an Asian guy before, and you're really cute" she told me candidly, a statement that revealed more about the cultural dynamics of the time than any sociological study could. In the complex social hierarchy of early 1990s military life, such observations were commonplace, though their implications ran deeper than most acknowledged. Our brief relationship became an exercise in navigating cultural curiosity and genuine connection.

Communication during this period was an adventure in itself. This was the era before cell phones, when reaching someone required strategic use of the barracks duty phone. Kimberly developed a system, introducing herself as my fiancée to bypass the duty desk's screening, on the phone and in person, a small deception that spoke to the creative ways young Marines maintained their social lives within military constraints.

The relationship burned bright but brief, lasting barely a month before her ex-boyfriend's return signaled its end. The situation felt familiar, echoing my high school relationship with Jenny. But this time, the pain was different. Where Jenny's rejection had once pushed me toward the Marine Corps as an escape, Kimberly's departure simply confirmed what the Corps had taught me about moving forward.

I then started taking weekend trips back to Riverside, a one-hour drive away, and it became escapes of a different kind, chances to step away from the intensity of base life and Oceanside and reconnect with a calmer reality and past.

These weekend trips became increasingly significant, offering not just escape from base life but connection to a community that understood me on a deeper level. The Korean church community, which had been part of my life since childhood, provided a different kind of social interaction, one rooted in cultural understanding and shared experience.

It was during one of these church gatherings that I met Kathy. Her status as an international student, with limited English proficiency, created an immediate point of connection. Our early conversations mixed my decent Korean with her emerging English, creating a unique dialogue that bridged cultural and linguistic gaps.

The complexity of her situation, studying in America on an F1 visa - added an unexpected dimension to our relationship. Our discussions quickly evolved beyond casual dating into more practical considerations. The idea of marriage, which might have seemed premature in other circumstances, emerged as a potential solution to multiple challenges.

For Kathy, marriage offered a path to permanent residency, though it meant potentially alienating her traditional parents in South Korea. For me, it presented an escape from the barracks lifestyle and access to additional military benefits. But beyond these practical considerations, we found in each other a kind of understanding that transcended our initial pragmatic approach.

The decision to marry Kathy came with its own set of complicated dynamics, particularly within the Marine Corps culture. Captain Murney's reaction epitomized the complex, almost paradoxical nature of military leadership, simultaneously rigid and deeply personal. The infamous line, "If the Marine Corps wanted you to have a wife, it would have issued you one," was delivered with her characteristic bluntness, yet beneath the harsh exterior lay a genuine concern for my well-being.

When I discussed the details of my relationship with her, Captain Murney's response was characteristically direct. She immediately declared that the relationship "didn't pass the smell test", a phrase that captured her intuitive sense that something about the marriage felt expedient rather than genuine. Her objection wasn't rooted in simple disapproval, but in a deeper understanding of the potential complexities of a marriage that seemed more administrative than emotional.

Yet, in the same breath, she made it clear that while she could offer her professional opinion, my personal life was ultimately my own domain. "This is your life," she told me. "While I don't agree with your decision, I respect that it's yours to make." Real leadership: setting clear expectations while ultimately trusting the individual to make their own choices.

Vegas

Our wedding was a perfect reflection of our relationship, a blend of practicality and unexpected romance. The Marine Corps 96, four days, 96 hours of liberty that felt like a lifetime of possibility, became our unexpected pathway to matrimony. One Thursday afternoon in May 1994, we made a spontaneous yet carefully calculated decision to drive to Las Vegas. The Stardust Hotel became our first shared home, a temporary haven of possibility and potential.

The next morning, we worked through the bureaucratic dance of love, obtaining our marriage license at the Clark County office with the same methodical precision I'd learned in military supply management. The Little White Chapel's drive-up window was less a romantic ideal and more a strategic opportunity, we were taking full advantage of the active duty free marriage benefit, a practical approach that seemed to embody the very nature of our relationship.

Our wedding reception was delightfully unconventional, McDonald's replaced the traditional banquet hall, a quick meal punctuating our impromptu ceremony. From there, we ventured to a casino, testing our luck with the same spirit of adventure that had brought us together. The day was a mixture of calculated decision and wild spontaneity, much like our relationship itself, practical yet filled with an unexpected sense of possibility.

Maintaining the secrecy of our marriage from Kathy's church community and parents required careful management. We presented ourselves as dating while privately dealing with immigration paperwork and military bureaucracy. Catholic Charities assisted in working through the legal procedures required by the INS to adjust status from an F1 visa to Permanent Resident through marriage. The dual life we led reflected the complex nature of our situation, a marriage born of pragmatism but growing into something deeper.

My mother's reaction stemmed from her own complicated past. Having been previously married to a man who had used citizenship marriage as a pathway to America, she viewed our relationship through the lens of her own personal trauma. Her warning, "Break up with her and find someone else who's a citizen", carried the weight of her own painful experiences.

The situation reached a crisis point in November when my mother, instead of confronting us directly, contacted Kathy's parents in Korea. The resulting chaos forced us to flee our Moreno Valley apartment one weekend, packing our belongings under cover of darkness. The next day, we learned that a man sent by Kathy's father had come looking for her. In that moment, her, and my nonexistent, relationship with her mother and father ceased.

Sergeant Sweetsir helped during this crisis. Through his connections, we found an apartment in Temecula adjacent to his, creating a fresh start away from the complicated family dynamics that threatened to derail us. His support extended beyond professional mentorship to genuine human compassion, demonstrating the deep bonds that could form within the Marine Corps community.

In Temecula, we began building a life that balanced military service with **personal** growth. The deployments to Kuwait and potential conflict in Saudi Arabia had awakened something in me, a recognition that military service was a means to an end, not a final destination. I enrolled in night classes at the local community college, taking the first steps toward a broader vision of our future.

Our marriage, which had begun as a practical arrangement, evolved into a partnership of mutual growth and support. Each challenge, from family disapproval and practical disownment to military obligations - strengthened our bond, proving that sometimes the most meaningful relationships grow from unexpected beginnings.

The military benefits of marriage brought practical improvements to our life. Moving out of the barracks and receiving BAH (Basic Allowance for Housing) allowed us to create a real home, something neither of us had truly experienced before. Our apartment, though modest, became a sanctuary from both military demands and family pressures.

Kathy's own growth was remarkable. Free from the constraints of her student visa and the expectations of her traditional Korean family, she began to discover her own identity. She made friends in our apartment complex and gradually built a life that was neither purely Korean nor American, but uniquely her own.

The cultural dynamics of our relationship became an unexpected asset in my military career. My ability to navigate between Korean and American cultures, already valuable in my technical role, gained new depth through our daily interactions. We created our own hybrid culture at home, blending military precision with Korean traditions and American practicality.

Captain Murney, who had initially expressed skepticism about our marriage, gradually came to respect the stability it brought to my life. She noticed how my performance improved with the structure of married life, how the chaos of barracks living had been replaced by a more focused approach to both work and personal development.

The balance between military duties and married life required constant adjustment. When deployment orders came, Kathy faced her first real test as a military spouse. Her isolation was compounded by her limited English and disconnection from family support, but she showed remarkable resilience. She threw herself into English classes and built a network of friends among other military spouses, many of whom were also making their own cultural transitions.

Our communication during deployments became an exercise in bridging multiple gaps, not just distance, but language and culture as well. Letters mixed Korean and English, creating a private language that reflected our unique relationship. Each deployment strengthened our bond rather than weakening it, as we both grew more independent while remaining connected.

The technical demands of my role in the SMU actually benefited from my more stable home life. The mental energy once spent managing barracks social life could now be directed toward system innovations and professional development. Kathy understood the long hours required for special projects, having grown up in a Korean culture that valued professional dedication.

Sergeant Sweetsir's influence in our lives continued even after his honorable discharge from the Marine Corps in early 1995. Still living in Temecula with his wife Maki, he remained a constant presence and mentor, though his focus had shifted from military leadership to entrepreneurial ambitions. Our friendship deepened beyond the military hierarchy that had initially defined it.

Mike, as we now called him post-discharge, introduced us to various business opportunities, starting with Amway. His enthusiasm for multilevel marketing and other get-rich-quick schemes was infectious, though not always profitable. We attended meetings and presentations together,

learning about business models and sales techniques that, while ultimately not our path, provided valuable lessons in critical thinking and
economic reality.

These evening meetings and weekend business seminars became a
regular part of our social life. Mike's constant stream of new business
ideas and opportunities reflected a common theme among veterans, the
search for purpose and success in civilian life. While we approached these
ventures with cautious skepticism, the time spent with Mike and Maki
strengthened our friendship and provided a bridge between military and
civilian worlds.

Our life in Temecula developed a rhythm that balanced military obligations with these entrepreneurial adventures. The apartment near
Mike's became more than a living space; it was our first real home together, a place where we could build our relationship away from the pressures of family expectations and military social life.

PREPARING FOR DEPLOYMENT

The technical innovations we'd developed in the SMU began to attract
attention beyond Camp Pendleton. As 1994 approached, two distinct
opportunities emerged that would test everything we'd built, first in Kuwait, then in response to growing tensions with Iraq. These deployments
became defining moments in both my military career and personal life.

Operation Native Fury in Kuwait would be our first major test, selected specifically because of our unit's technical capabilities. But even as
we prepared for this deployment, none of us could have predicted that
we would return to the region just months later under the shadow of
potential combat operations during Operation Vigilant Warrior.

My personal life added new complexity to these preparations. Kathy
and I had just begun dating, and the prospect of multiple deployments

would test our developing relationship. The cultural expectations she carried as a Korean international student didn't always align with the demands of military life, but these challenges would ultimately strengthen our bond.

As we prepared for Kuwait, Captain Murney's words stuck with me: "This is why we innovate, not for the exercises we can predict, but for the contingencies we can't." Her wisdom proved prophetic as events unfolded through 1994, and everything we'd developed at Camp Pendleton would be put to the ultimate test.

The months ahead would challenge everything we thought we knew about our capabilities, both technical and personal. But for now, we focused on the immediate task at hand, preparing for Kuwait, not knowing it would be just the first act in a much larger story.

CHAPTER 5

KUWAIT

110600ZMAR94

Two deployment options were announced during a SMU morning formation in February 1994: Somalia or Kuwait. One was a UN humanitarian mission in a country descending into chaos. The other was a joint exercise called Operation Native Fury, testing the systems and personnel that would respond if Saddam Hussein moved on Kuwait again. I chose Kuwait, not because it was safer, but because it offered something Somalia couldn't: a chance to prove that the barcode tracking system we'd developed could revolutionize military logistics under real-world conditions. What I didn't know was that the deployment would give me far more than a line on my service record. It would give me a legendary stopover in Ireland, a haunting reminder of war's cost in an abandoned schoolhouse, and a feast under the stars with a grateful sheikh. This was where I stopped being a supply clerk and started being a supply Marine.

THE CHOICE

In February 1994, two distinct deployment opportunities presented themselves to the Marines of Camp Pendleton's SASSY Management Unit. Operation Restore Hope offered humanitarian logistics in Somalia. Operation Native Fury in Kuwait focused on testing maritime prepositioning capabilities in the still-tense Persian Gulf.

While many Marines were drawn to the higher-profile Somalia mission, I approached the choice analytically. Kuwait represented a structured environment where technical innovations could be properly tested. The maritime prepositioning exercise would involve tracking thousands of pieces of equipment through complex supply chains; an ideal proving ground for the barcode system I'd been developing.

Captain Murney supported my reasoning. "This isn't just about moving equipment," she explained. "It's about proving that innovative supply management can transform how we handle large-scale operations."

The preparation phase required navigating security protocols as much as technical challenges. Accessing the Maritime Prepositioning Force ship manifests demanded a comprehensive clearance. Weeks of background investigations followed: interviews with family members, reference checks, deep background reviews. The MPF manifests weren't just equipment lists; they were blueprints of military preparedness, everything from tank components to individual spare parts. When the clearance finally came through, it validated not just my technical expertise but my trustworthiness with sensitive logistics.

The technical preparation consumed every remaining hour. Kuwait's heat and sand could destroy electronic equipment. Power supply issues, connectivity challenges, hardware durability, each potential failure point required backup procedures. We tested the system against worst-case scenarios, subjecting equipment to extreme temperatures and deliberately introducing errors to verify integrity.

The night before deployment, I found myself in the SMU office running final system checks. Captain Murney appeared, something rare for that hour. "You've chosen an ambitious path," she observed. "But that's exactly what we need, Marines who see beyond conventional limitations."

THE BARCODE REVOLUTION

The barcode tracking system emerged from a fundamental understanding of military logistics challenges. Maritime Prepositioning Ships functioned as floating warehouses carrying everything from tanks to toilet paper, but existing tracking methods relied on manual processes that were time-consuming and error-prone.

Our core innovation was a distributed database system that could function both online and offline. Each piece of equipment received a unique barcode containing serial number, service life data, maintenance requirements, and unit assignment. But the real breakthrough was how we structured the data relationships. Tracking an M1 tank wasn't just about monitoring a single piece of equipment. The tank consisted of thousands of components, each with its own maintenance schedule. Our system created a digital tree structure where scanning the tank's primary barcode instantly revealed the status of all subordinate parts.

The barcode scanners needed to be rugged enough for Kuwait's harsh environment. We developed protective cases and cooling procedures. Power management became crucial; the system had to function reliably with intermittent power supplies. We built a multi-tiered backup that could store data locally and sync when connections became available.

Training adapted to different skill levels. Basic users learned simple scanning and data entry. Advanced users learned system administration and troubleshooting. The interface used familiar military terminology and straightforward procedures. Marines in field conditions couldn't navigate complex menus or technical jargon.

Working with Sergeant Hill, we refined deployment procedures. His practical experience bridged the gap between technical capabilities and real-world requirements. We developed rapid setup procedures that could have the system operational within hours of arrival.

THE JOURNEY

The deployment began at San Diego International Airport, where a chartered Tower Air flight awaited 150 Marines. Standing on the tarmac in desert utilities, weapons cleaned and stored in hard cases, we represented the forward element of Operation Native Fury. The morning marine layer created a misty backdrop that made the massive aircraft seem almost ethereal.

Shannon

The flight plan included a refueling stop in Shannon, Ireland, a detail that would become legendary. The planning staff had scheduled a twenty-minute ground stop, seemingly unaware of the implications of letting 150 Marines loose in an airport famous for its duty-free shops and bars.

What followed was a masterclass in Marine Corps initiative. Within minutes, the pub was packed with Marines racing to order both a Guinness and an Irish coffee: a combination that seemed perfectly reasonable in our jet-lagged state. The bartenders, clearly veterans of military charters, moved with practiced efficiency.

Then came an announcement that elevated the stop to mythical status: the aircraft required additional maintenance, extending our ground time by an hour. The cheer that went up from 150 Marines, all still carrying their weapons (though unloaded), must have startled the regular travelers. Looking back, the wisdom of allowing armed military person-nel to get thoroughly intoxicated seems questionable, but such were the relaxed protocols of the early 1990s.

By the time we reboarded, the mood had shifted dramatically. What had begun as a tense deployment turned into something approaching a flying party. The remainder of the flight passed in a haze of altitude-en-hanced intoxication and fitful sleep.

THE ABANDONED SCHOOL

Kuwait International Airport presented a stark contrast to Shannon's jo-vial atmosphere. Stepping off the plane into pre-dawn darkness, we were immediately struck by the heat. It was an oppressive wall that seemed to steal breath from our lungs. Even at that early hour, the temperature ex-ceeded 90 degrees.

Our destination was an abandoned school compound near Al Shuaiba Port, chosen for its tactical advantages, a single entry point with water to the rear made it easily defensible. But "abandoned schoolhouse" doesn't capture what we found. This was a war-torn building, a ghost of education past. Bullet holes pockmarked the exterior walls, some clus-tered in patterns suggesting sustained firefights. Shrapnel scarring radi-ated outward from blast points. Entire sections of wall had been blown apart.

The first order of business was cleaning. All of us, from boot Lance Corporals to senior NCOs, worked together to make this place habitable. We swept debris, patched what we could, established defensive positions.

The reality of being a boot Lance Corporal hit immediately. While senior Marines organized the command center, I found myself assigned to managing the burn pits, dragging out half-drums and establishing the waste management system. Twenty Marines were assigned to each for-mer classroom. Our room was dominated by the constant playing of

Confederate Railroad, courtesy of Corporal Bennett, a die-hard West Virginia native whose boom box seemed to have endless battery life. The music became a strange comfort, a slice of Americana in the middle of the Arabian desert.

The Student ID Cards

It was during the cleanup that my team made a discovery that haunts me still. In one of the destroyed classrooms, student ID cards lay scattered across the floor. Small laminated rectangles with young faces, names in Arabic script, birthdates that made them children when the war came through.

We gathered them carefully, unsure what else to do. No one spoke. There was nothing to say. These cards represented students who should have been learning mathematics and literature, not fleeing from violence that had torn their school apart. What happened to them? We would never know.

The cards went into a box, and we finished our work in silence.

THE SYSTEM MEETS REALITY

Setting up the technical infrastructure proved challenging in these conditions. The heat affected equipment in ways we hadn't fully anticipated despite our testing. We learned to perform most computer work in early morning or late evening hours. The school's electrical system required careful management to prevent overloads.

The true test began with the arrival of the Maritime Prepositioning Ships at Al Shuaiba Port. These massive vessels carried entire ecosystems of military equipment—everything a Marine brigade would need for thirty days of sustained operations. The complexity of tracking thousands of items would push our system to its limits.

Our operation center, established in a former science classroom, hummed with computers and cooling fans fighting against the Kuwait heat. When the initial wave of equipment began flowing from the ships, the sheer volume threatened to overwhelm our processing capacity. We implemented a triage approach, categorizing equipment by priority and complexity.

The barcode system proved its worth immediately. Traditional tracking would have required hours of manual documentation; our scanners processed items in seconds. Each scan captured not just inventory data but maintenance requirements, unit assignments, and location tracking.

Desert heat and fine sand threatened scanner operations constantly. We developed rotation systems allowing equipment to cool down while maintaining continuous operations. The protective cases proved their worth, though we modified some components to improve heat dissipation.

The offline capability we'd built proved crucial. Port operations often took us out of network range, but local data storage and synchronization ensured continuous tracking. Each scanner functioned as a self-contained database that could operate independently and merge data when connectivity was restored.

Perhaps the most significant validation came from the units we supported. Field commanders began relying on our system's real-time equipment status updates. The transparency and accuracy built confidence in the supply chain, allowing units to plan operations with greater certainty about resource availability.

THE SHEIKH'S FAREWELL

At the end of the exercise, we received an unexpected gift. A local sheikh, grateful for the American presence and perhaps curious about these young men and women who had occupied the abandoned school, purchased steaks for the entire deployment. Not just any steaks, but thick, quality cuts that seemed impossibly luxurious after weeks of MREs and field rations.

Over a hundred Marines gathered for what became an impromptu feast. The smell of grilling meat filled the compound, mixing with the ever-present dust and diesel. We ate until we couldn't move, then found room for more. Some Marines were on their third or fourth steak before they finally surrendered.

It was a moment of pure, uncomplicated joy. For one evening, we weren't worried about guard duty or logistics systems or the geopolitical tensions that had brought us here. We were just young people, far from home, sharing a meal that someone had generously provided. The sheikh's gift wasn't just food; it was a gesture of connection, a reminder that even in places scarred by war, hospitality and humanity persist.

The deployment's success caught the attention of higher command. As a Lance Corporal, I was put up for, and then awarded, a Navy Achievement Medal, an award typically reserved for more senior ranks. The recognition challenged the traditional notion that junior Marines couldn't make significant contributions to large-scale operations.

The abandoned school in Kuwait, the student ID cards, the barcodes, the impromptu bar in Shannon, the sheikh's steaks: these were more than memories. They were markers of growth, evidence of the complex, often unexpected path of military service.

Kuwait had been practice. I didn't know it yet, but the real test was coming.

CHAPTER 6

VIGILANT WARRIOR

210600ZOCT94

Seven months after Kuwait, Saddam Hussein moved his Republican Guard toward the Kuwaiti border again. This time it wasn't an exercise. On October 8, 1994, the alert came down through the chain of command, and the machinery of war began grinding into motion. Within weeks, I found myself in Saudi Arabia, standing guard duty at a port near Jubail, weapon at the ready, watching headlights approach in the darkness and wondering if this was the moment everything changed. Kuwait had been practice. Saudi Arabia was the real thing, or close enough that the difference stopped mattering. I wrote letters to Kathy during those long night watches, careful words that didn't say how scared I was, how much I wanted to come home, how thin the line felt between peace and something else entirely. This was the deployment that taught me what almost-war feels like, and how its weight stays with you even after the threat dissolves.

THE ALERT

October 8, 1994, arrived with the weight of gathering storm clouds. Intelligence confirmed significant Iraqi Republican Guard movements toward the Kuwaiti border. Within hours, Camp Pendleton transformed from its regular operational tempo to a heightened state of readiness. For those of us selected for the FSSG Forward Element, the alert triggered an immediate shift in priorities: create a mobile, self-contained supply infrastructure capable of supporting combat operations in potentially hostile territory.

The technical preparations consumed every waking moment. Drawing from lessons learned during Operation Native Fury, I needed to develop supply systems that could function without constant connectivity; completely self-contained operations that could process transactions offline and batch upload whenever satellite communication became available. Each system had to be tested and retested. In a combat environment, supply chain disruptions could have lethal consequences.

My marriage was barely six months old. Kathy and I had discussions we never expected to have so early: power of attorney, emergency contacts, worst-case scenarios. Her Korean family struggled to understand the realities of military life. Their concept of marriage didn't easily accommodate the possibility of long separation under dangerous conditions.

Thirty Minutes

By October 22, the alert status escalated to active standby. The restrictions were immediate: no one in the Forward detachment could venture beyond a thirty-minute drive from base. We had to remain constantly contactable, our lives suspended in perpetual readiness. Some Marines had to temporarily relocate closer to Camp Pendleton.

I was fortunate. Our apartment in Temecula sat precisely at the thirty-minute limit, allowing some semblance of normal life during this period of heightened tension. But "normal" had lost its meaning. Every evening with Kathy felt borrowed. Every morning could be the last before deployment orders came.

The uncertainty was the hardest part. We operated in constant readiness, never knowing if each day would bring deployment orders or stand-down instructions. This limbo affected everything, personal plans, daily routines, the ability to think beyond the next few hours. The weight of potential combat operations hung over every moment.

As October drew to a close, the tension reached new heights. Each intelligence update brought fresh concerns about Iraqi intentions. Our technical systems were ready, our procedures refined, our equipment prepared. All that remained was the waiting.

MARCH AIR FORCE BASE

October 27, 1994, brought the green light we had both anticipated and dreaded. Lines of commercial buses waited to transport us to March Air Force Base, diesel engines idling in pre-dawn darkness. The military caravan north was marked by unusual quiet. Marines who typically filled such transitions with boisterous conversation and dark humor now sat in contemplative silence. This wasn't a training exercise. It was a direct response to aggressive military movements by a recently defeated but still dangerous adversary.

Fourteen hours stretched before us at March Air Force Base, a marathon of anticipation. Some Marines tried to maintain composure through endless equipment checks. Others attempted to sleep on uncomfortable terminal seats. I oscillated between reviewing technical procedures and quiet reflection. The supply systems I had prepared were more than logistical tools now; they would be critical infrastructure supporting potential combat operations.

The C-130 Hercules aircraft waited on the tarmac like patient beasts of burden. As day turned to dusk, their propellers occasionally turned in the evening breeze. These aircraft would carry us into uncertainty, toward a situation that could escalate into combat at any moment.

My thoughts drifted to Kathy, to our apartment in Temecula, to the life we had begun building together. The conversations of the past weeks played through my mind, her support mixed with understandable worry, our discussions about the unpredictability of military life.

The flight itself became an exercise in endurance. Seated in cargo nets along the aircraft's interior, we felt every turbulent bump, every course correction. The noise made conversation nearly impossible, forcing each of us into our own thoughts.

INTO THE DESERT

The C-130 carved through Middle Eastern airspace, pushing us toward Dhahran, Saudi Arabia. Through small windows, we caught glimpses of the Arabian landscape below. It was an endless expanse of desert that would soon become our operational theater.

Our arrival procedures reflected the heightened tensions. The aircraft executed a tactical approach, defensive maneuvers designed to minimize vulnerability to potential ground threats. Even in these early moments, the reality pressed upon us: this wasn't just another training exercise.

The difference between this deployment and Kuwait hit me within the first few hours. Kuwait had been an exercise, a chance to test systems and build practices. We'd worked hard, but there was always an underlying sense that we could make mistakes, learn from them, adjust.

Saudi Arabia was execution. There was no room for error.

Everything we'd done in Kuwait, we now had to do with military precision and in less than half the time. The offloading of maritime prepositioned ships, the scanning and tracking of equipment, the distribution to units, it all happened at a pace that would have seemed impossible during Native Fury. My barcode system, which had been an innovation in Kuwait, became absolutely critical here. Without the speed and accuracy it provided, we would have been drowning in manual paperwork while the clock ticked down.

Setting Up Camp

General Purpose tents arose from the desert floor in precise rows, each designated for specific functions. The supply operations center required particular attention because our technical systems needed to be operational within hours of arrival.

The first forty-eight hours tested every aspect of our preparation. Power generation became an immediate challenge; our equipment required stable electrical supply in an environment hostile to mechanical operation. The desert heat demanded constant adjustment to our cooling systems. Fine sand penetrated every seal and connection.

As a newly promoted Corporal, I found myself not just operating technical systems but leading others in their operation. The skills learned from Kuwait helped, but new challenges emerged daily. Each morning brought fresh technical obstacles: sand-clogged filters, overheating equipment, communication disruptions requiring immediate solutions.

FEAR IN THE DARKNESS

If Kuwait's guard duty had been tense, Saudi Arabia's was something else entirely. Our bivouac site near the port felt exposed, vulnerable. The possibility of bomb-laden vehicles wasn't an abstract fear here, it felt imminent, real. Every approaching engine noise triggered a surge of adrenaline.

I found myself with my weapon at the ready multiple times during guard shifts. Suspicious activity. Vehicles slowing near our perimeter. Headlights that seemed to be surveilling our position. Figures moving in patterns that didn't match civilian behavior. Each time, I had to make split-second assessments. Was this a threat? Was I about to be the first line of defense against an attack that could kill everyone I served with?

The nights stretched forever.

I developed a habit during those long watches: I wrote letters home. Sitting in the darkness, weapon across my lap, I'd pull out paper and pen and write to Kathy by the dim light available. The letters were careful. I didn't want to worry her more than necessary. But they were honest about missing home, about hoping to return, about the uncertainty that hung over everything.

Some nights, I found myself praying. Not the formal prayers of church services, but the desperate, wordless prayers of someone who wants very badly to survive. I prayed that the suspicious vehicles were just civilian traffic. I prayed that the next shift would arrive without incident. I prayed that this whole deployment would end peacefully and I'd make it back to California, back to the life we were trying to build together.

The other Marines on guard duty felt it too. We didn't talk about it directly, that wasn't how Marines operated. But you could see it in the way everyone stayed extra alert, extra focused. This wasn't training anymore.

THE SANDSTORM

The technical systems faced their greatest test when a major sandstorm hit our position. For three days, we operated in conditions that pushed both equipment and personnel to their limits. Visibility dropped to mere feet. Fine sand penetrated every conceivable seal and filter. Yet our supply operations couldn't stop; combat readiness depended on our ability to maintain accurate tracking and distribution of resources.

We implemented emergency protocols we'd developed but hoped never to use. Operating in shifts, teams worked continuously to protect equipment while maintaining essential services. We sealed sensitive equipment in makeshift environmental chambers created from plastic sheeting and air filters, improvising solutions with whatever materials we could find.

During the storm, the stress showed in different ways across the unit. Some Marines threw themselves into physical training when off duty. Others became unusually quiet or irritable. The combination of technical demands, harsh environment, and combat readiness requirements took its toll on everyone.

When the storm finally passed, each piece of equipment told its own story of adaptation and endurance. As we cleaned sand from components and documented wear patterns, we were cataloging lessons that would inform future deployments.

STANDING DOWN

Then, almost as suddenly as the crisis had begun, diplomatic channels reported movement toward resolution. Saddam Hussein's decision to withdraw troops from the Kuwaiti border came like a collective exhale rippling through our forward position. What had seemed an inevitable march toward conflict dissolved with unexpected speed.

The standing-down process proved as methodical as our deployment. Equipment that had been rapidly deployed now needed to be carefully preserved and prepared for transport. The desert environment had taken its toll. Sand had infiltrated nearly every piece of equipment, extreme heat had stressed mechanical components in unexpected ways.

The process of standing down revealed unexpected emotional challenges. After weeks of maintaining peak readiness, the sudden shift toward departure created a strange mix of relief and anticlimax. Marines who had been tightly focused on potential combat operations now had to redirect their energy toward methodical withdrawal.

Looking back at our deployment area as we prepared to leave, I understood that something fundamental had changed. We had stood at the precipice of conflict and pulled back. The technical systems, the supply chains, the organizational procedures, everything we had prepared had contributed to a show of force that ultimately helped prevent conflict. We had proven that our systems could function under extreme conditions, that our training could translate into real-world effectiveness.

THE WEIGHT OF ALMOST-WAR

March Air Force Base materialized from the California darkness, a welcome sight after Saudi Arabia. The familiar landscape of Southern California seemed somehow different, as if the lens through which we viewed the world had been permanently altered.

The impact on my marriage revealed itself in subtle ways during those first weeks home. Kathy and I had weathered our first significant test of military life. Our conversations carefully danced around the deployment, acknowledging its weight without dwelling on the anxiety we had both endured. Seven months into our marriage, we had already faced the possibility of combat deployment. It was a reality check that prompted serious discussions about our future.

A latent form of stress began manifesting in unexpected ways. Unlike dramatic combat stress often portrayed in movies, this was quieter, more insidious. The knowledge of how close we had come to actual combat created a heightened awareness that was difficult to switch off. Every news report about the Middle East triggered a cascade of tactical thinking, equipment readiness, operational procedures.

Sleep patterns took weeks to normalize. The hypervigilance developed during deployment didn't simply switch off. Many of us found ourselves automatically scanning for threats, maintaining unnecessary operational security, carrying the weight of combat readiness into our civilian interactions.

During this period, I made a quiet but significant decision: while I would continue to serve with total commitment, I needed to prepare for life beyond the Marine Corps. The deployment had shown me both the importance of military service and its potential costs. I enrolled in more college classes, focusing on technical subjects that would open doors beyond the military.

The deployment had transformed us from technical specialists into battle-tested professionals, even though the battle had never actually commenced. The experience became a turning point, though its full impact wouldn't be clear until years later. The technical expertise, leadership experience, and understanding of large-scale operations would shape my future career. But more importantly, Saudi Arabia had taught me something about myself, and about my ability to perform under pressure, to innovate in challenging conditions, to lead others in situations of genuine uncertainty.

And it taught me what almost-war feels like. How the weight of it stays with you. How you carry the letters you wrote and the prayers you whispered long after the threat dissolves and you're home again, trying to remember what normal used to mean.

CHAPTER 7

KANEOHE BAY

011600ZMAY96

By 1996, my upcoming rotation presented only two options: Camp But-
ler in Okinawa or Marine Corps Base Hawaii. Sergeant Sweetsir's stories
about Okinawa were enticing, but Hawaii offered something more. A
chance for genuine reinvention that could include my professional ambitions and my marriage. The ability to bring Kathy, continue my education, and serve in a strategic location made Kaneohe Bay the clear choice.

The separation from Camp Pendleton on March 27, 1996, exactly
three years to the day I had first arrived as a boot PFC, carried symbolic
weight. The goodbye to Mike Sweetsir, now a civilian after his honorable
discharge, marked a transition from military mentorship to genuine
friendship. Three years earlier, he'd been the sergeant who showed me
how things actually worked. Now we were peers, and I was heading to
Hawaii with skills and confidence I couldn't have imagined as that nervous lance corporal checking into the SMU.

ARRIVAL IN PARADISE

We landed at 1600 on May 1, 1996. Honolulu International Airport hit us with warm humidity and the smell of plumeria before we even got off the plane. Our sponsor, Sergeant Jeremy Cole, waited in khakis and an aloha shirt, a combination that would have gotten you laughed out of Camp Pendleton. He draped traditional leis over our necks and helped us load our bags into his minivan.

The drive to Kaneohe took us over the Likelike Highway, and I remember Kathy grabbing my arm when the mountains came into view. Lush green peaks erupting straight up from the earth, shrouded in mist, with waterfalls cutting through the vegetation. After three years of Camp Pendleton's dusty brown hills, it looked like another planet. I understood immediately why Marines fought for this duty station.

The base's Temporary Lodging Facility was full, so the Marine Corps put us up at the Hilton Hawaiian Village for twenty-five days. Government-paid luxury while we waited for housing. I kept expecting someone to realize there'd been a mistake, but nobody ever did. We ate breakfast overlooking Waikiki Beach and tried to remember we were still in the military.

Checking into Combat Service Support Group 3 dissolved any remaining tension. The command climate was different here. People actually seemed relaxed. My experience at Camp Pendleton's SMU served as an immediate passport to respect. The commanding officer had heard about the barcode system we'd implemented in Kuwait and wanted to know if we could do something similar for Pacific theater operations.

Within a few months of arrival, Kathy got pregnant. The locals we met just nodded knowingly. "Hawaiian water," they said, like it was the most obvious thing in the world. Apparently, this happened to a lot of couples who transferred here. Something about the islands.

SMU HAWAII

The SMU at CSSG-3 represented the cutting edge of military logistics technology placed in an island paradise. My assignment to the Operations office provided an unprecedented opportunity to push boundaries. Hawaii's sophisticated telecommunications infrastructure allowed us to experiment with emerging technologies that would have been impossible at Camp Pendleton.

The promotion to Sergeant brought new dimensions to my role. Leadership responsibilities now extended beyond system development to include training and mentoring other Marines. Captain Murney's lessons helped as I balanced these dual responsibilities.

Our most significant achievement came in developing a distributed logistics management system that could operate across the Pacific theater. The system had to account for unique challenges of operating across multiple nations, each with its own regulations. Working with counterparts in Japan, Korea, and other Pacific nations, we developed interfaces that bridged different military logistics systems while respecting national security requirements.

The success of our innovations led to increased visibility. Our unit became a testing ground for new Marine Corps logistics concepts, with teams from Headquarters Marine Corps frequently visiting to study our methods. What started as solutions to specific Pacific theater challenges evolved into templates for military-wide logistics improvements.

ANABEL

Tripler Army Medical Center sits on a hill overlooking Pearl Harbor, a pink building that looks more like a resort hotel than a military hospital. I've heard it's one of the largest military hospitals in the world. On May 1, 1997, it was just the place where my daughter decided to arrive.

Kathy had been in labor for hours by the time I got there. Supply Marines don't get paternity leave the way civilian fathers do—I'd been working until the last possible moment, then rushing across the island hoping I hadn't missed everything.

I hadn't.

The delivery room was smaller than I expected, filled with more people than seemed necessary. Doctors, nurses, someone whose job I never quite figured out. Kathy was exhausted but focused, doing the work that only she could do.

When Anabel came out, she pooped immediately.

This is the detail that has stuck with me for twenty-five years. Not the crying, not the weight and length measurements, not even the first moment of holding her. She arrived in this world and immediately made a mess. Everyone laughed. The tension broke. Welcome to parenthood.

"Look at all that hair," one of the nurses said.

She was right. Anabel had more hair than any newborn I'd ever seen—a full head of dark hair that made her look like a tiny adult. The nurses kept commenting on it as they cleaned her up.

They handed me scissors to cut the cord. The moment felt ceremonial but also strangely clinical. This was a medical procedure, a separation that had to happen, and they were letting me participate. I made the cut and felt something shift—some new weight of responsibility that I wouldn't fully understand for years.

Later, the doctor asked if I wanted to help examine the placenta.

"We have to make sure it's complete," he explained. "Any fragments left behind can cause complications."

So I found myself digging through biological material with a doctor I'd just met, checking for completeness, making sure my wife would be okay. It was disgusting. It was fascinating. The doctor asked if I'd ever considered medicine.

"No," I said. "I'm a supply guy."

I washed my hands probably fifteen times before holding my daughter again. She was seven pounds, eleven ounces, something ounces—the numbers have faded. What hasn't faded is the weight of her in my arms, the impossibility of being responsible for something so small, so dependent, so entirely new.

I was twenty-four years old. I had no idea what I was doing.

SALT LAKE

Military base housing operated as a rigidly controlled environment where housing managers conducted surprise inspections with the intensity of combat readiness assessments. Regulations governed everything from grass height to outdoor furniture placement.

My most memorable encounter came during a routine lawn mower rental. After returning the equipment in what I considered acceptable condition, the retired Sergeant Major running the MWR facility delivered a masterclass in military verbal discipline. His tirade transformed a simple lawn mower return into a full-dress inspection failure. That incident proved to be our breaking point.

By May 1998, with our daughter Anabel just turning one year old, we moved to a seventh-floor apartment in Salt Lake, Honolulu. The Lee family became our anchor to this new community, first-generation Korean-Americans who ran a local convenience store that felt like a portal to my childhood memories. Their teenage children reminded me of my own upbringing working at my family's pizza shop. Here is where I found a love for Spam Musubi.

The Basic Allowance for Housing more than covered our modest 650-square-foot apartment in Hawaii's expensive market, with enough left over for a financial cushion. Through the Lee family's store and neighborhood connections, we discovered local markets, family-owned restaurants, and a Korean church that would have remained unknown in base housing.

The commute along the newly built H-3 highway, with its stunning views of the Ko'olau mountains, provided natural transition between military and civilian life. Several neighbors worked in tech, and conversations expanded my understanding of civilian applications for military technical skills, insights that improved my approach to military systems.

GLENN KNEPP

Our first encounter with Sergeant Glenn Knepp came from a moment of domestic crisis. Newly moved into base housing, we found ourselves locked out. Knepp's solution revealed his Recon background immediately: scaling our second-floor balcony with fluid grace, he unlocked our door from the inside.

"What exactly is your MOS?" I asked.

"Recon," he replied with characteristic understatement, as if climbing buildings was simply part of an average day.

Together, we launched a small side business rebuilding and selling PCs, modeling our operation on companies like Dell and Gateway. The venture combined Knepp's strategic thinking with my technical expertise. We implemented supply chain management techniques from military logistics, approached customer service with the same dedication we brought to our Marine Corps duties, and learned that military skills could translate effectively into civilian success.

Knepp's imminent transition to civilian life became a case study for my own future planning. Watching him methodically prepare, researching industries, networking with potential employers, translating military skills into civilian terms, taught me a lot. "The civilian world," he would say, "is just another terrain to master. You need to understand the lay of the land, identify your objectives, and plan your approach accordingly."

Nearly thirty years later, our friendship endures. While we some-times go years without talking, we remain always just a phone call away, able to reconnect and dive into conversations about business ideas as if no time had passed. What began with a locked door became a bond that transcended our time in Hawaii.

LORETTA CORNETT-HUFF

Anabel's birth became a catalyst for change. Hawaii Pacific University's Computer Science evening program offered a concrete path toward the future I'd envisioned when joining the Marine Corps. My daily routine became a carefully orchestrated balance: classes from 1730 to 2130, homework until midnight, then up at 0600 for unit PT. This intensity stood in sharp contrast to my earlier Camp Pendleton days.

The technical coursework proved surprisingly complementary to my military duties. Database concepts learned in evening classes im-proved my work in the SASSY Management Unit the next morning. Pro-gramming languages (Java, C++, HTML, JavaScript) became my new tools, but the underlying principles remained: systematic problem-solv-ing, attention to detail, creating functional systems.

Ms. Loretta Cornett-Huff, the Joint Education Services Officer, be-came my pivotal mentor. With over 42 years of government service, she saw potential where others saw obstacles. Through her guidance, I learned to strategically challenge courses and leverage testing opportuni-ties. She introduced me to the Rotary Club, expanding my professional network beyond military circles. I became the education center's "poster child", completing my Bachelor of Science in Computer Science in just two years instead of four.

The relationship extended beyond professional mentorship into genuine friendship with Loretta and her husband Paul. One memorable Thanksgiving at their Kailua condo, we spent hours searching for their pot-bellied pig Snowball, whom Loretta feared had been captured for a traditional Hawaiian Kalua feast. When Snowball wandered home around 9 PM, completely unaware of the concern she'd caused, the relief was palpable. The incident exemplified the unique blend of professional mentorship and personal friendship that characterized our relationship.

PREPARING FOR KOREA

The announcement of Operation Foal Eagle in late 1998 arrived at a pivotal moment in my Marine Corps career. With nearly two years at Kaneohe Bay, I had established myself as a technical leader within CSSG-3, while simultaneously progressing through my computer science degree. The deployment would test not just my military capabilities, but the full spectrum of skills I'd developed in Hawaii.

Preparation for the exercise demonstrated how far I'd come from my early days at Camp Pendleton. The technical systems we'd developed at Kaneohe Bay needed adaptation for deployment conditions, requiring innovations that combined my academic knowledge with practical military experience. Each solution had to account for both technical requirements and the realities of operating in Korea's challenging environment.

My role as a Korean linguist added another dimension to the preparation. The language skills maintained through family connections and formal study would now serve a critical military purpose. This deployment would mark the first time my cultural background, technical expertise, and military training would fully align in service of a mission.

Staff Sergeant Leonard Hill became my primary collaborator in developing our deployment capabilities. Together, we worked to create systems that could merge transactional data from multiple supply units over satellite communication, a capability that proved essential during the exercise.

The weeks before deployment revealed the strength of the community we'd built in Hawaii. The Knepp family helped look after Kathy and Anabel, demonstrating how military friendships extend beyond individual service members to encompass entire families. The educational groundwork laid with Ms. Cornett-Huff ensured my academic progress wouldn't be completely disrupted by the deployment.

This deployment would prove different from my previous experiences in Kuwait and Saudi Arabia. Where those had been primarily technical challenges, Operation Foal Eagle would demand a broader range of capabilities. The preparation period reflected this complexity, requiring attention to linguistic and cultural considerations alongside technical readiness.

The technical preparation became increasingly focused on cross-unit compatibility. Our systems needed to interface with both American and Korean military networks, adding layers of complexity to our usual deployment preparations.

Personal preparation took on new dimensions. While reviewing technical manuals and checking equipment, I also found myself refreshing Korean language skills that would be crucial for our mission. The prospect of serving as both technical specialist and interpreter created both excitement and anxiety, this deployment would test every aspect of my military and personal development.

As departure day approached, I found myself better prepared than ever before for a deployment. The professional growth fostered at Kaneohe Bay, combined with academic progress and strong family support, created a foundation for success that paid off in Korea. The upcoming experience would represent not just another deployment, but a unique opportunity to demonstrate the full spectrum of capabilities developed during my time in Hawaii.

CHAPTER 8

KOREA

030600ZOCT98

I had never been to Korea. My grandmother spoke the language to me as a child, and I'd passed the interpreter exam at the highest level, but the country itself existed only in family stories and Saturday language school memories. When Operation Foal Eagle sent me to Pohang in October 1998, it was supposed to be a joint exercise testing our supply systems. It turned into something else. For six weeks I walked the land my mother and grandmother had left behind, ate food I'd only known from her kitchen, and discovered that the name I'd reclaimed at that recruiting office in Riverside, the name my biological father gave me before disappearing, actually meant something here. Staff Sergeant Leonard Hill and I built systems that worked. The ROK Marines taught me things about my own heritage I'd never understood. And somewhere between the frozen nights and the farewell dinner, I figured out that being caught between two cultures wasn't a weakness. It was the whole point.

HOMECOMING

Operation Foal Eagle was more than another military exercise. For me, it was a confluence of personal history, technical work, and cultural identity that would reshape my understanding of both my military service and my heritage.

The Korean language had been the soundtrack of my childhood, woven into my earliest memories. My grandmother, a formidable woman who spoke only Korean, had been my first and most demanding language instructor. Our conversations were a delicate dance of patience and persistence, her limited English matching my limited Korean. Those childhood lessons were far more than simple language training. They were cultural preservation, a bridge to my roots that I didn't fully appreciate until much later.

Formal Korean language education had been a non-negotiable part of my upbringing. Weekend Korean language schools were common for children of immigrant families, but for me, they carried extra weight. I learned to read and write Korean with precision. The Korean interpreter exam at Level 4, the highest level, was less a challenge than a confirmation of skills honed through years of family immersion.

When the opportunity for Operation Foal Eagle came up, my involvement seemed almost predetermined. I was uniquely positioned: a Marine with deep technical expertise in supply operations, a secondary MOS as a Korean language interpreter, and a personal connection to the culture that ran deeper than most could understand. The remote supply operations system I had developed wasn't just a technical achievement. It could be a game-changer for military logistics in joint operations.

My personal motivations ran deeper than professional ambition. Despite my Korean heritage, I had never actually set foot in Korea. The deployment represented a pilgrimage of sorts, a chance to walk the land of my ancestors, to see the places that had only existed in family stories and childhood memories. It was an opportunity to connect with a part of my identity that had always existed at the edges of my understanding.

Staff Sergeant Hill understood this complexity. Our conversations about the deployment went beyond technical specifications. We discussed the challenges of international military operations, the balance between technological innovation and cultural understanding. He saw in me not just a technical specialist, but a bridge between cultures, between systems, between what was and what could be.

The flight to Korea felt significant from the moment we lifted off. Hill and I spent much of it fine-tuning our remote supply operations system. Military flights are rarely quiet, but our corner of the aircraft became a mobile command center. Laptops open, satellite communication protocols being reviewed, we were already bridging the gap between preparation and operational readiness. His expertise in logistics complemented my technical background perfectly. Where he saw supply chain challenges, I saw algorithmic solutions waiting to be implemented.

CAMP MUJUK

Pohang International Airport emerged from the landscape like a gateway between my two worlds. The cold hit first, sharp and unforgiving, a stark contrast to Hawaii's warmth. As we deplaned, my senses were overwhelmed. The sounds, the smells, the very air felt different. This was Korea not as a distant family story, but as a living, breathing reality.

Camp Mujuk would be our home for the duration of Operation Foal Eagle. The base was a study in military efficiency, positioned strategically near the Port of Pohang. Our initial setup was spartan: General Purpose tents that would become our entire world for the next few weeks. The cold was relentless, seeping through every layer of clothing.

Setting up our supply operations center was a delicate dance of technological precision. Our remote capabilities system was a carefully constructed network that would allow multiple supply units to merge their transactional data over satellite communication. Hill and I worked with a synchronicity that spoke to weeks of preparation, our movements almost choreographed as we established our technological infrastructure.

My role as both supply operations chief and Korean language interpreter became immediately apparent. Where most Marines saw obstacles, I saw opportunities for connection. The local Korean workers became bridges to understanding, to culture, to a part of myself I was only beginning to fully explore.

The cold was a constant companion. Our GP tents, equipped with a small potbelly heater, became a lifeline. It was here that I began to bring in small touches of local culture: fresh chestnuts to roast, finding ways to boost morale that went beyond standard military protocols. Hill would joke that I was turning our supply operations center into a "home away from home," but he appreciated how we could make the harsh environment more bearable.

THE ROK MARINES

A pivotal moment came during a joint training exercise with the Republic of Korea Marines. The day-long session became much more than standard military training. It evolved into real cultural exchange. As a Korean-American sergeant, I found myself in a unique position, bridging not just technical systems but cultural understanding between the two forces.

The ROK Marines showed particular interest in my background, curious about the path that had taken me from Korean heritage to the U.S. Marine Corps. Our conversations, conducted in a mix of Korean and English, went beyond military matters to explore themes of identity, duty, and cultural adaptation. They were especially intrigued by how I had maintained my Korean language skills while growing up in America, and how this cultural fluency enhanced my military service.

The respect from the ROK Marines was multifaceted. They appreciated both my technical expertise and my cultural understanding. As a sergeant who could communicate fluently in their language while managing complex military systems, I represented a bridge between traditional Korean values and modern military capabilities. Our discussions often turned to how military service was viewed in both cultures, creating deeper bonds between our units.

These interactions led to relationships that transcended the typical temporary nature of joint exercises. The ROK Marines' interest in my story opened doors for deeper cooperation. During our joint training sessions, what began as professional military exchange often evolved into discussions about the intersection of duty, family, and cultural identity. They were particularly fascinated by how I balanced American military service with Korean cultural values. "You understand both worlds," one ROK Marine sergeant told me. "This makes you stronger, not divided."

Technical briefings became exercises in cultural diplomacy. When explaining our supply systems to ROK counterparts, I could switch seamlessly between Korean and English, ensuring that complex technical concepts were clearly understood. This ability earned respect from both American and Korean leadership.

The cold of Pohang seemed to foster a unique camaraderie. During breaks between operations, ROK and U.S. Marines would share cigarettes, hot coffee, and military stories, with me often serving as both translator and cultural interpreter. These informal moments, where military protocol relaxed slightly in favor of human connection, proved as valuable as our formal training sessions.

LEONARD HILL

SSgt Leonard Hill and I had spent months developing our remote supply operations system, and now it would face its ultimate test. The system we'd created could merge transactional data from multiple supply units over satellite communication. Where previous deployments relied on physical data transfer and manual reconciliation, our system promised real-time integration of supply information.

The technical infrastructure was a delicate ecosystem. Mainframe computers, satellite communication equipment, and custom-developed software created a network that could process and merge supply transactions instantaneously. I watched as other supply units arrived, their data systems interfacing with our central hub. Each successful merge was a small victory, a validation of the many hours of preparation.

Our satellite communication capabilities were impressive for that era. We had created a system that could seamlessly transfer complex supply data across vast distances. Hill and I would exchange knowing glances with each successful data merge, understanding the significance of what we were achieving.

The cold weather presented unexpected technical challenges. Equipment that had performed flawlessly in Hawaii's climate needed constant adjustment in Pohang's winter conditions. We developed solutions for keeping our systems operational, from makeshift environmental controls to new backup procedures. Each challenge became an opportunity to demonstrate the adaptability of our approach.

The collaborative nature of our work went beyond professional cooperation. Hill had become more than a colleague. He was a partner in a complex technological and cultural mission. Our conversations would drift from technical specifications to broader discussions about military logistics, about how technology could transform military operations. He respected my dual capabilities in a way that went beyond typical military hierarchies.

116

Local procurement became another fascinating aspect of our operation. My language skills allowed us to work with local supply chains with efficiency that would have been impossible for other units. We weren't just requesting supplies. We were building relationships, understanding local logistics in ways that transcended typical military procurement.

Hill's logistics expertise combined with my technical and cultural knowledge created an effective problem-solving team. When issues arose, we could approach them from multiple angles: technical, operational, and cultural. This comprehensive approach became our trademark.

The most significant test came when a three-day communication blackout hit during a severe winter storm. The offline processing capabilities we'd built in allowed operations to continue uninterrupted, with data automatically syncing once connections were restored. This reliability caught the attention of higher commands, who saw potential for applying similar approaches across other military systems.

By the end of the exercise, we had processed thousands of transactions, coordinated complex supply chains across multiple units, and maintained seamless communication between U.S. and ROK forces. As we packed up equipment and prepared for departure, Hill summed up our achievement: "We didn't just build a supply system. We created a new way of thinking about military logistics."

MR. KIM

The Korean driver assigned to our unit was named Mr. Kim. Of course he was. In Korea, roughly one in five people share the surname, so the odds were always in favor of a Kim. But this Mr. Kim became essential to our operation in ways that went far beyond his official duties.

He drove a beat-up military surplus truck that looked like it had survived the Korean War, and maybe it had. His job was to help us access local supplies, run errands, ferry materials around the Pohang port area. On paper, it was a simple logistics support role. In practice, Mr. Kim became our fixer, our cultural guide, and eventually something like a friend.

I got to know him during the long supply runs around the port. We talked in Korean, which surprised him at first. He kept looking over at me while he drove, like he was trying to figure out the puzzle. A U.S. Marine sergeant, clearly Korean by blood, speaking his language fluently, but admitting this was his first time ever setting foot in Korea. He found this fascinating and slightly absurd.

"Your Korean is better than some Koreans," he told me once, shaking his head. "But you hold your chopsticks like an American."

He wasn't wrong.

Mr. Kim could get us anything. Building supplies, materials for the operations center, outside vendors for equipment we couldn't source through official channels. He knew everyone, it seemed, or at least knew someone who knew someone. The Korean networks of connection and obligation that I'd heard about my whole life, but never really understood, suddenly made sense watching him work. A phone call here, a conversation there, and somehow the impossible became possible.

The contraband was his specialty. Captain Q whiskey appeared regularly, sourced from somewhere Mr. Kim never fully explained. He'd just show up with a few bottles, accept whatever payment we offered, and move on to the next request. I learned not to ask too many questions.

But my favorite thing Mr. Kim ever got us was a bag of fresh chestnuts.

It was deep into the deployment, cold as hell, and morale was starting to drag the way it always does when you've been living in tents for weeks. Mr. Kim showed up one afternoon with this burlap sack, maybe ten pounds of chestnuts, still in their spiky outer shells. He handed it to me like it was nothing special.

The smell hit me immediately. My grandmother used to make these. She'd boil them in a big pot on the stove, and the whole house would fill with that earthy, sweet smell. I hadn't thought about it in years, but suddenly I was eight years old again, sitting in her kitchen, waiting for them to cool enough to eat.

We didn't have a pot, but we had coffee cans. Large ones, scrounged from the mess tent. And we had the potbelly diesel heater that kept our GP tent from becoming a freezer. So we filled the coffee cans with water, dropped in the chestnuts, and balanced them on top of the heater to boil.

It took a while. The heater wasn't exactly designed for cooking. But eventually the water started bubbling, and that smell filled the tent. Marines started wandering over, curious about what we were doing. Most of them had never had fresh chestnuts before. Some had only seen them in that Christmas song and didn't realize they were an actual food people ate.

We pulled them out of the water, let them cool just enough to handle, and showed everyone how to peel them. The shells come off easier when they're hot, but you have to be careful not to burn your fingers. The inner skin is the tricky part. Get that off and you've got this soft, sweet, slightly nutty meat inside.

I watched Marines who'd grown up on farms in Iowa and housing projects in Baltimore and suburbs in California all sitting around our supply tent, eating boiled chestnuts and making the same surprised face when they tasted them. "These are actually good," one lance corporal said, like he'd expected a trick.

Yeah. They were.

Mr. Kim brought us other things too. Korean sweets and snacks that most Americans had never seen. Rice cakes, dried squid, these little honey cookies that disappeared within minutes of arriving. We'd share them around, and it became a way of introducing the other Marines to a culture most of them knew nothing about. Food is the easiest bridge.

119

People might be skeptical of customs or language or religion, but nobody turns down something that tastes good.

The MREs became our currency. Being the primary supply unit, I had access to essentially unlimited Meals Ready to Eat. They were worth more than cash in certain situations. The Korean workers loved them, partly for the novelty and partly because some of the components were genuinely useful. The instant coffee packets, the candy, the crackers with cheese spread. We traded MREs for supplies, for services, for information. I once got a critical piece of equipment repaired in exchange for a case of MREs and a handshake.

Mr. Kim thought this was hilarious. "Americans throw away what Koreans would keep," he said. "Koreans throw away what Americans would keep. Trade is just finding the difference."

He was a philosopher, Mr. Kim. In his beat-up truck, running supplies around Pohang, he'd figured out something about how the world worked that most people never learn.

By the end of the deployment, he'd become more than just our assigned driver. He was part of the team. When we packed up to leave, he came by to say goodbye, and I could tell he was genuinely sad to see us go. We'd been good business, sure, but I think it was more than that. For a few weeks, he'd been part of something, connected to these Americans who'd landed in his country and actually tried to understand it.

I gave him my watch as a thank-you gift. It wasn't expensive, just a standard military field watch, but he acted like I'd handed him something precious. He said something in Korean that roughly translates to "we'll meet again," though we both knew we probably wouldn't.

I still think about Mr. Kim sometimes. I hope he's doing well. I hope he found other Americans worth helping, other opportunities to be the guy who could get things done. Korea is full of Mr. Kims, I'm sure. But that one was ours.

LIBERTY

Liberty calls during military deployments are sacred moments, brief windows of freedom that become lifelines amid the structured intensity of military operations. In Pohang, these moments became something more. They were opportunities for cultural discovery, for pushing beyond the boundaries of typical military experience.

My role as a translator transformed these liberty calls into extraordinary excursions. Where most Marines might have been limited to tourist traps or base-adjacent establishments, I became their cultural guide, their key to a world they would never have experienced otherwise. Each excursion was carefully planned, not just a tour, but an immersive experience designed to challenge preconceptions and open minds.

We explored markets that tourists rarely see, navigating the complex alleyways of Pohang with a mix of my language skills and the Marines' adventurous spirit. Street food became cultural exploration. Spicy tteokbokki, savory fish cakes, local vendors' special recipes. These became living lessons in Korean culture. I translated not just words, but the stories behind the food, the cultural significance of each dish.

The local bars and restaurants became unexpected classrooms. I would translate not just language, but context, explaining the social rules of Korean drinking culture, the significance of how one pours a drink, the importance of respect in social interactions. Captain Q and other local liquors became keys to understanding social dynamics. Soju became a love-hate relationship for all of us.

Morale became my unexpected specialty. I worked with local drivers and support staff to bring in small comforts that transformed our spartan deployment environment. Fresh chestnuts roasted over our potbelly heater became a nightly ritual. I negotiated for local delicacies, found ways to source Captain Q, a local liquor that became a prized commodity among the Marines. These weren't just supplies. They were connections, small bridges between two cultures.

The Bathhouse

The traditional bathhouse visit became legendary among our unit. I'll never forget the collective shock of the Marines as they discovered the true nature of a Korean bathhouse: completely nude, a cultural practice that challenged every notion of personal space and privacy they had brought from America. Some guys tried to keep their towels on at first. The Korean patrons just stared at them like they were the weird ones. Which, in that context, they were.

The stories that emerged from that single outing would be told and retold for years. Marines who had been to combat zones, who had faced down every kind of physical challenge the Corps could throw at them, reduced to awkward shuffling because they had to sit in a hot tub naked with strangers. It was beautiful.

The Juicy Bar

One particularly risky but memorable evening led to an unexpected connection across generations of Marines. A small group of us, including a Master Sergeant Riggs, visited what was known as a "juicy bar," establishments that were part of military lore in Korea. What made this visit remarkable was Riggs' revelation that he had frequented this same establishment over twenty years earlier during his first deployment.

The situation became surreal when we discovered that the current madame in charge was someone Riggs had known during his earlier deployments, now running the establishment she had once worked in as a young hostess. The recognition was immediate and created an awkward moment of cross-generational military history that I now had to manage as translator.

As the designated interpreter, I found myself in the uncomfortable position of translating conversations that bridged not just language but decades of shared history between American military presence and local establishments. The madame shared stories about how Pohang had changed, how the military presence had evolved, and how she had transitioned from hostess to business owner. Translating these narratives while maintaining appropriate professional distance required careful judgment.

The evening became a strange history lesson about the long-standing relationship between U.S. Marines and the local community. Through their conversation, which I carefully translated, emerged stories of how these establishments had adapted over the decades, how relationships between Marines and locals had evolved, and how some things remained surprisingly unchanged despite the passing years.

For the younger Marines present, it was an eye-opening glimpse into a different era of military-civilian relations in Korea. For me, it was an exercise in diplomatic translation, carefully managing the cultural and generational gaps while maintaining everyone's dignity. Some things, I learned, were better left partially translated.

THE NAME I RECLAIMED

My own relationship with Korea was complex and deeply personal. Despite my heritage, I was experiencing the country as much as my fellow Marines were: as a visitor, albeit one with a deep ancestral connection. Each interaction, each translation, each cultural explanation was also self-discovery. I was learning about my homeland through the eyes of my fellow Marines, seeing it with a perspective that was both insider and outsider.

The local population's response to our presence was nuanced. They saw beyond the uniform, recognizing in me a different kind of connection. My ability to speak Korean, to understand not just the language but the cultural subtleties, created rapport that went far beyond typical military-civilian interactions.

The name I'd reclaimed at that recruiter's desk in Riverside, the name my biological father had given me before disappearing, now meant something in a country I'd never visited but had always carried inside me.

One particularly meaningful interaction occurred when we visited a local temple. The monks, impressed by a Korean-American Marine who could explain Buddhist concepts to his fellow servicemembers, invited our group for tea. What could have been a simple tourist visit became a meaningful discussion about spirituality, duty, and the intersection of different worldviews. The Marines, sitting cross-legged on temple floors, drinking green tea and asking thoughtful questions, demonstrated how genuine curiosity could transcend cultural barriers.

As the deployment progressed, I realized that my secondary MOS as an interpreter was perhaps my most valuable contribution. The technical systems we developed were important, but these human connections, these moments of genuine cultural understanding, were truly transformative. I was not just a Marine, not just a supply operations specialist, but a cultural ambassador, a living bridge between two worlds.

The evening meals became opportunities for deeper connection. I would explain the significance of different dishes, the proper way to eat certain foods, the cultural importance of sharing meals. Marines who had never used chopsticks found themselves mastering the skill, encouraged by friendly local restaurant owners who appreciated their earnest attempts to respect Korean customs.

FAREWELL

The ROK Marines organized a farewell dinner that exemplified the bonds we'd built. The evening mixed military formality with Korean traditions, creating an atmosphere where both cultures were celebrated. Senior ROK officers spoke about how our technical innovations had been enhanced by cultural understanding, while U.S. commanders acknowledged how the deployment had transformed their perspective on international cooperation.

During the dinner, a ROK Marine major pulled me aside. "You've shown us something important," he said in Korean. "That being between two cultures isn't a weakness. It's a power that can make both sides stronger." His words captured something fundamental about what we'd accomplished during Operation Foal Eagle.

The packing up process became a time of reflection. Each piece of equipment we carefully stored represented not just technical capability but memories of challenges overcome, relationships built, lessons learned. Hill and I worked methodically, ensuring everything was properly documented while sharing our thoughts about what we'd achieved.

The flight back to Hawaii offered time to process the experience. The technical systems we'd developed had exceeded expectations, but the human connections we'd forged felt equally significant. We had demonstrated that military effectiveness in the modern world required both technical excellence and cultural intelligence.

Landing at Kaneohe Bay felt like returning to a different world. The warm Hawaiian air was a sharp contrast to Pohang's winter chill, but the biggest adjustment was shifting from an environment where my cultural background had been a daily asset back to regular military routine. The deployment had permanently altered how others in the unit viewed my capabilities.

The success of our technical innovations led to increased responsibilities. Captain Murney, having received glowing reports about our system's performance, immediately tasked us with implementing similar innovations for Pacific-wide operations.

The deployment had also changed things at home. My time in Korea had given me a deeper appreciation for Kathy's own experience as a

Korean in America. Our conversations about cultural identity took on new depth. I understood something now that I hadn't before.

Professional recognition followed our return. The technical achievements during Operation Foal Eagle became a case study in effective international military cooperation. I found myself briefing senior officers about not just the technical aspects of our system, but the importance of cultural understanding in military operations.

Hill and I continued to refine our systems, incorporating lessons learned from Korea. "We're not just improving technology," he said during one of our work sessions. "We're changing how the Marine Corps thinks about international operations."

The Korea deployment was over. But standing in the Kaneohe Bay supply center, running the same systems we'd pioneered in Pohang, I knew something had shifted. For the first time in my Marine Corps career, I started thinking seriously about what came next.

CHAPTER 9

TRANSITION

140800ZMAY99

The Korea deployment changed something in me. Not just the technical achievements or the cultural reconnection, but a sense that I'd done what I came to the Marine Corps to do. My Computer Science degree was nearly complete. I had seven years of service, international deployments, technical systems that actually worked. The restless kid from Riverside who joined to escape had become someone with actual skills, actual accomplishments, actual options. Kathy had spent another month managing everything alone. The question wasn't whether I could stay. The question was whether staying was the right move. The Corps had given me everything I'd asked for. Another enlistment would give me more of the same, and more deployments, more years of my daughter growing up with a father who came and went. If I could land a civilian job before my reenlistment decision date in February 1999, I'd have my answer.

Loretta's Last Lesson

Loretta Cornett-Huff had guided me through four years of accelerated education. Now she was going to help me translate all of it into something civilian employers could understand.

"David," she said during one of our final meetings, "your military experience is a treasure trove. We just need to help civilian employers understand its value."

We sat in her office at the education center, my resume spread between us, and she taught me a new language. Supply chain management became "logistics optimization." Technical system implementations became "IT infrastructure overhauls." The barcode system I'd built in Kuwait became a "process improvement initiative that reduced inventory discrepancy rates." Same work, different words.

I sent out over a hundred applications. Most disappeared into whatever void job applications go to die. But twenty led to phone interviews, and five of those led to something more.

The Harvey Hotel

The Raytheon interview was different from everything else.

Most of my other interviews happened in Hawaii. Companies flew recruiters out, or we did phone screens, or in a few cases I met people during brief trips to the mainland. Professional, straightforward, nothing remarkable.

Raytheon flew me to Dallas.

When I landed, I discovered why. The Harvey Hotel had been essentially taken over by Raytheon for a mass hiring event. Hundreds of candidates, maybe more, all there for the same reason: Raytheon had just acquired TI Defense and E-Systems the year before and made the decision to consolidate all defense manufacturing into Tucson. Existing employees got a choice between relocation or severance packages. Only about thirty percent took the relocation.

Which meant Raytheon needed to hire over three thousand new engineers. Fast.

The first evening was a sales pitch. Convention-style seating, presentations about Raytheon's programs, why we should want to work there. They were selling us as much as we were selling ourselves. I sat in a ballroom full of engineers and computer scientists, most of them with graduate degrees and years of industry experience, wondering if my seven years of Marine Corps logistics would even register.

The next day was the interviews. Speed dating for engineers.

My resume had been circulated to all the hiring teams beforehand, and five teams had selected me to interview. The format was unlike anything I'd experience later, nothing like Amazon's Loop process. Each team got exactly fifty minutes. They came to my hotel room, usually two people: the hiring manager and a senior team member. They sat across from me in the cramped hotel furniture, asked their questions, took their notes, and left. Then the next team arrived.

Fifty minutes to decide if they wanted to make an offer. Fifty minutes for me to decide if I wanted to work with them.

Three of the five teams made offers. I picked Camille Trotter's team. The work sounded interesting, and they needed someone with a security clearance, which I already had from the military. One less hurdle.

I flew back to Hawaii not knowing for certain. The call came a few days later from Jim Pearson, the recruiter. Senior Systems Engineer I. Start date May 21, 1999. Salary that made my sergeant's pay look like pocket change.

THE OFFERS

The numbers still surprise me when I think about them. Raytheon, General Dynamics IT, Microsoft. Offers ranging from $40,000 to $60,000. As a sergeant in 1999, my base pay was $19,369.56 a year. These offers were closer to what a colonel made.

Jim Pearson made the difference at Raytheon. Retired Navy commander, he understood what seven years of military service actually meant. The Senior Software Engineer I position typically required seven to ten years of post-college experience. I didn't have that. What I had was seven years of increasingly complex logistics operations, international deployments, and systems I'd built from scratch under conditions that would make most corporate IT departments quit.

Jim saw that. He translated it the same way Loretta had taught me to translate it. Military experience counted.

I accepted the Raytheon offer. Start date one week after my planned discharge.

SEPARATION

May 1, 1999. Exactly three years to the day after arriving in Hawaii, we began our departure. The symmetry wasn't lost on me. Military life had a way of creating these almost poetic moments of mathematical precision.

Marine Corps Base Hawaii couldn't process separations, so we flew to Los Angeles and drove to Camp Pendleton. The Temporary Lodging Facility became home for two weeks while the Marine Corps methodically dismantled my identity as a Marine.

Exit interviews. Separation physicals. Administrative procedures. Each station, medical, dental, finance, was another piece being catalogued and filed away. Hurry up and wait, the phrase that had defined seven years, was never more applicable.

I would miss the sounds. Reveille in the morning, taps at evening, the ceremonial raising and lowering of the flag. These weren't just sounds. They were a rhythm that had defined my existence. The Separations Company at Mainside was surreal. I moved and acted as a Marine, knowing that in days I would no longer be one.

May 14, 1999. Honorable discharge. No grand ceremony. Just a piece of paper, the DD-214, that officially made me a civilian. The transition was both instantaneous and impossibly gradual.

First Day

The orientation room at Raytheon's Spring Creek campus in Dallas looked like a college lecture hall, except everyone held manila folders full of paperwork. Some people were literally holding framed diplomas.

I didn't have a diploma. Not yet.

When I'd accepted the offer months earlier, the assumption was simple: I would complete my degree before discharge. But a disagreement with my capstone course professor had derailed everything. After years of near-perfect academic performance, a GPA close to 4.0, I'd failed the capstone. Not because I couldn't do the work. Because the professor and I had fundamentally different views on how software should be built.

I'd filed a complaint with the Dean of Computer Science. He reviewed my record, granted a waiver, and let me retake the course with a different instructor. The problem was timing. The retake wouldn't finish until November. My first day at Raytheon was May 21.

I started sweating before I sat down.

"Welcome to Raytheon," the HR coordinator announced. "We'll begin with credential verification. Please have your official transcripts and degree certificates ready."

The line formed. Bachelor's in Electrical Engineering. Master's in Computer Science. Ph.D. in Physics. The credentials stacked up like a poker game where everyone had better cards than me.

When I reached the verification table, my hands were damp.

"Degree certificate?" the administrator asked without looking up.

"I've got this one."

I turned to find Jim Pearson, silver-haired, with bearing that said military before he opened his mouth. "Come with me, son."

He led me to a quieter corner. My mind raced through options. Maybe I could explain. Maybe I'd be driving back to my apartment to figure out what the hell to do with my life.

"You're the Marine, right? Supply MOS?"

"Yes, sir." The sir came automatically.

Jim smiled. "Relax. I'm retired Navy, not your commanding officer." He pulled out a folder with my name. "You're an experienced hire. Different track. Your military logistics experience counts in lieu of the degree requirement. Degree expected by end of year?"

"November, sir. Capstone situation, but the dean approved a retake."

Jim nodded like professors and capstone disputes were problems he'd seen before. "Get it done. What we don't know won't hurt us."

By the end of orientation I had a red badge, a cubicle assignment, and a start date. The military network had followed me into civilian life.

Camille Trotter found me in the break room later.

"You're David Kim, right? The Marine?"

"Yes, ma'am."

She laughed. "It's Camille. And you can drop the ma'am. You're not in the military anymore."

But in some ways, I was. And it would take years to figure out which parts of that identity to keep and which to leave behind.

THE CAPSTONE

Six months later, I completed the retaken capstone course with a 100%. Perfect score. The same approach that earned me a failing grade with one professor earned me an A+ with another.

The project was straightforward: a Java-based inventory management system demonstrating mastery of data structures, algorithms, and database integration. The conflict wasn't about whether I could build it. It was about how.

The first professor wanted traditional Software Development Life Cycle. Complete requirements documentation up front. Detailed design specifications. Architecture diagrams. Everything approved before writing a single line of code. Waterfall methodology, the industry standard at the time.

I didn't work that way. I built the application iteratively. Started with a core function, tested it, refined it, added the next piece. The database schema evolved as I discovered what the application actually needed. Documentation came after the code worked, not before. I delivered a fully functional system that met every technical requirement, but my process documentation was thin because I'd been building and testing rather than planning and specifying.

The professor's verdict: I couldn't possibly have developed a working solution without following proper SDLC. Therefore I must have cheated or cut corners. Failing grade.

What I'd done, without knowing it had a name, was practice iterative development. Build, test, refine, repeat. Years later I'd learn the term "agile" and realize I'd been instinctively following its core values before the Agile Manifesto was even written.

The second professor evaluated the deliverable. Working application. Clean code. Every required competency demonstrated. The fact that I'd built it through iteration rather than waterfall didn't diminish its quality.

Perfect score. Degree conferred. The chapter that had caused me more stress than any deployment was finally closed.

CHRISTIAN

Christian was born November 13, 1999, at 8:55 in the morning. Tucson Medical Center. And we almost didn't make it.

Kathy's water broke early that morning at our place in Oro Valley. I helped her to the car, got her settled, started the engine, and looked at the gas gauge.

Empty.

Not low. Empty. The needle was below the E.

So there I was, pulling into a gas station while my wife screamed through contractions in the passenger seat. Contractions that were coming fast. Too fast. I pumped gas as quickly as humanly possible while she gripped the dashboard and made sounds I'd never heard another human make. I genuinely thought we might deliver that baby in the car.

We didn't. Barely.

I got her to the hospital and into a wheelchair. The nurses rushed her to delivery. Her doctor was paged but hadn't arrived yet. Hadn't even had time to scrub in. I remember seeing her running down the hallway, still drying her hands.

She didn't make it in time either.

The nurse caught Christian as he came out. Kathy was trying not to push because they kept telling her not to push, but Christian had other plans. He was coming whether anyone was ready or not. The nurse, this young woman who looked almost as surprised as we were, caught him before he could hit the floor.

Twenty minutes. That's how long we'd been in the hospital. Twenty minutes from wheelchair to delivery.

The nurse was overjoyed. It was the first baby she'd ever "delivered" by herself, if you count catching a baby who's determined to arrive as delivering. Her hands were shaking when she handed him to the doctor, but she was smiling like she'd won something.

Christian was healthy. Loud. Impatient from his very first moment in the world, unwilling to wait for anyone's timeline but his own.

I stood there watching them clean him up, this second child I hadn't planned for but suddenly couldn't imagine life without. Anabel was two and a half. Now she had a brother. I had a son.

Two kids. A new job. A degree finally completed. The MBA program starting soon. I was twenty-five years old and building a life that looked nothing like where I'd started.

THE IDENTITY QUESTION

The hardest part of civilian transition wasn't learning new skills or adapting to corporate culture. It was reconciling who I had been with who I was becoming.

For seven years, "Marine" wasn't just my job. It was my entire identity. The uniform, the discipline, the sense of mission. These weren't aspects of employment. They were foundational elements of who I was.

In those first months at Raytheon, I experienced something I'd later recognize as identity grief. Walking into the office each morning without the weight of a uniform, without the clear chain of command, without the shared understanding of purpose that permeated every Marine Corps space, I felt unmoored. The rituals that had structured my existence for seven years, morning formations, military courtesy, the discipline of precise timing, had vanished overnight.

Who am I if not a Marine?

The question sat with me silently. Technical skills translated easily enough. The deeper sense of belonging to something greater than myself was harder to replace.

Kathy noticed before I did. "You still stand at parade rest while waiting in lines," she said one weekend at a mall. "And you still wake up at 0500 even though you don't need to."

Unconscious habits. Comfort rituals. Muscle memory for an identity I was no longer living but hadn't fully released.

The breakthrough came gradually. Not in abandoning my Marine identity but in expanding it. I wasn't leaving the Corps behind. I was bringing its core values, integrity, attention to detail, commitment to excellence, into a new arena. The mission had changed. The warrior remained.

THE ROAD AHEAD

Standing in my cubicle at Raytheon, red badge on my lanyard, I couldn't have imagined what the next two decades would bring.

The military taught me to plan meticulously, to anticipate obstacles, to prepare for contingencies. But life doesn't follow operational orders. The career I was building would take me through corporate promotions and lateral moves, from Raytheon to consulting firms to positions I couldn't yet envision. I would experience success that exceeded my wildest expectations and failures that would bring me to my knees.

There would be a restaurant called Yama that I'd pour my savings into, trusting a partner who would betray that trust. There would be a marriage that couldn't survive the weight of financial pressure and two people growing in different directions. There would be a bankruptcy filing that forced me to sign away everything I'd built.

And there would be rebuilding. Again.

The Marine Corps had prepared me for that, even if I didn't know it yet. The discipline to start over. The resilience to fail and get back up. The understanding that your lowest moment is not your final moment.

But those stories belong to a different chapter. On that spring day in 1999, walking out of Raytheon's campus after my first week, I was simply a twenty-seven-year-old veteran trying to figure out who he was without a uniform.

I had a job. I had a family. I had a degree finally finished and an MBA program on the horizon.

What I didn't have was any idea how much I still had to learn.

The Marines had forged me. The civilian world would test whether that forging would hold.

PART II

THE FULL ARC

When I first set out to write this book in 2020, it ended at Chapter 9. The story of a Marine who served his country, learned valuable lessons, and transitioned successfully into civilian life. A neat arc with a satisfying conclusion. Discharge papers signed, Raytheon badge issued, future secured.

That version was easier to write. But it was incomplete.

For years, I resisted telling the rest. The parts where discipline failed me. Where the same confidence that made me effective in the Corps made me reckless with money and relationships. Where the man who could manage supply chains across the Pacific couldn't manage his own household finances. Those chapters sat unwritten because they required admitting things I'd spent years trying to forget.

But a memoir that ends at the moment of apparent success isn't honest. It's a recruiting brochure.

What follows is not a story of unbroken success. It's a story of building, breaking, and building again. Of a restaurant that consumed my savings and my trust. Of a divorce that cost me my home and nearly my relationship with my children. Of a bankruptcy that forced me to sign away everything I'd spent decades building. And then, the rebuilding. A 400-square-foot apartment in Austin. Meeting Lacey. Learning what partnership actually looks like. Discovering that the lessons from the Corps weren't wrong, just incomplete.

The military taught me to complete the mission. Life taught me that sometimes the mission changes, and the real test is whether you can adapt.

This is the part of the story I almost didn't tell. I'm glad I changed my mind.

CHAPTER 10

THE FALL

210900ZMAY99

The fifteen years between Raytheon and bankruptcy contained everything. Corporate success, entrepreneurial failure, a marriage that couldn't survive the distance I put between us. I climbed higher than I'd ever imagined, then watched it all collapse. This chapter covers more ground than any other because that's how life works sometimes. The building and the breaking don't happen in separate, neat compartments. They overlap. They feed each other. The same ambition that drove the climb created the conditions for the fall.

I went from an engineer grateful for his first real salary to a Managing Director overseeing a regional consulting practice. I built a restaurant from the ground up, selected every stud in the walls, and watched men I trusted steal it from me. I maintained two households across two cities, telling myself the distance was temporary while my debt climbed and the

141

marriage slowly bled out. I earned more money than my parents ever dreamed of and managed to lose all of it.

The narrative that follows isn't a clean arc of rise and fall. It's messier than that. It's years of momentum in one direction, then another, decisions that seemed reasonable at the time and catastrophic in hindsight. I'm not proud of all of it. But I'm not hiding from it either. The man who rebuilt from bankruptcy could only exist because the man before him burned everything down. This is the story of that burning.

RAYTHEON

My first performance review at Raytheon became an unintentional cultural study.

After six months of exceeding targets and completing projects ahead of schedule, I walked into the meeting expecting what I'd gotten in the Marines: clear feedback, specific corrections, explicit guidance on improvement. That's how evaluation worked. You did well, they told you what you did well. You screwed up, they told you exactly how.

Instead, my manager seemed uncomfortable.

"Well, you're doing great overall," he hedged. "Maybe you could consider being more... politically flexible in some situations."

Political flexibility wasn't a metric I understood.

"Sir," I asked, falling back on military courtesy, "could you give me a specific example?"

His eyebrows rose slightly at the "sir." Another cultural misstep I hadn't recognized.

"Well, in the meeting last week, you disagreed with Ted's approach pretty directly. Ted is close to the VP, so sometimes it's better to find alignment with his ideas."

In the Marines, the best idea won regardless of rank. Diplomatic phrasing mattered, but truth prevailed. Here, professional feedback considered political alliances rather than objective merit. Corporate truth was situational. Relational.

I nodded and took notes like it was a tactical briefing. "Understood. I'll work on identifying key stakeholder relationships and adjusting communication accordingly."

What he didn't realize was that I was treating corporate politics like another foreign culture to be studied. Not so different from learning to work with ROK Marines or Kuwaiti officials.

That night, I created a relationship map of the organization's key players. Reporting lines, informal alliances, communication preferences. The same methodology I'd used to track supply chain relationships in military deployments. If corporate culture was the new terrain, I would map it with military precision.

YAMA

In April 2003, I left Raytheon to chase a dream that had been simmering since my childhood at Young's Pizza and Ribs.

The opportunity came through three men I'd met at church. A deacon. A pastor. A fellow congregant I'd called a friend. These weren't random business contacts from some chamber of commerce mixer. These were men of faith, men who prayed before meals and spoke of integrity from the pulpit.

That trust would prove to be my most expensive mistake.

We had a concept: fine dining Japanese restaurant in the Tucson foothills, near Canyon Ranch Spa. Wealthy clientele looking for high-end experiences. A gap in the market for authentic Japanese cuisine. Everything aligned.

We called it Yama.

Building that restaurant became an obsession. The chef and I hand-selected every stud that went into the walls, every stone that formed the entryway. This wasn't a franchise operation. This was craftsmanship, the same attention to detail I'd brought to rebuilding engines with Mr. Red-wine, the same precision I'd applied to supply systems in the Marine Corps.

A sushi bar that seated fifteen. Table seating for sixty. And our signature: a shabu shabu bar with eight seats, the first of its kind in Arizona. Each guest had their own pot of simmering broth, cooking thin slices of wagyu beef tableside. I personally tasted every bottle on our hundred-wine list. The sake menu alone took three months to develop.

I hired every staff member. Trained them personally. Learned to work every station because I refused to ask anyone to do something I couldn't do myself. The Marine Corps had taught me that.

Restaurant staff are a different breed. Young, attractive, living on tips and adrenaline. Our bartender had been voted hottest college bartender by the University of Arizona student body. And in that pressure-cooker environment, boundaries blur. Late nights closing together. Shared cigarettes in the alley. The walk-in cooler became a space where things happened. Or almost happened.

One afternoon, Rachel, a server, walked up to me at the sushi bar. Slim, redheaded, the kind of beautiful that made guests request her section. She leaned in close.

"I just found out I'm pregnant," she said. Then, without missing a beat: "That means we can totally go to the walk-in."

I didn't take her up on it. I wanted to. I'd be lying if I said otherwise. But something held me back. Maybe the last shred of the man I'd promised Kathy I'd be. Maybe just fear of complicating an already complicated situation. I said no, and she shrugged and went back to her tables like she'd offered me a stick of gum.

That was restaurant life. Temptation was part of the mise en place.

Yama opened to immediate success. Tucson foodies found us. Canyon Ranch guests discovered us. Within three months, we hit an annual revenue run rate over one million dollars. I started developing expansion plans. A second location. A quick-service version for the food court.

Then my partners held a board meeting without me.

I wasn't notified. Wasn't invited. Didn't know it happened until it was over. They voted to dilute my shares, restructure ownership in a way that pushed me out. When they presented the paperwork, they'd reduced my stake to almost nothing. Offered a buyback at pennies on the dollar.

The partnership documents gave them the power to do exactly what they'd done. Legal. Clean. Complete betrayal.

But this wasn't just business betrayal. This was betrayal by men who had broken bread with me at communion. Who had spoken of honesty and righteousness on Sundays, then executed a plan to steal my livelihood on Monday. They'd weaponized my faith against me, knowing I'd trusted them precisely because of where we'd met.

The weeks after Yama were the darkest of my life up to that point. Family depending on me. Mortgage. Bills. No income, no business, no partners. Only the bitter knowledge that I'd been outmaneuvered by people I'd called friends.

I'm not proud of what went through my mind during those weeks. There were nights I drove past the restaurant and imagined watching it burn. The satisfaction of seeing everything they'd taken reduced to ash. I never did it. What stopped me wasn't just morality. It was pragmatism. Burning it down would destroy me too. The math didn't work.

So I did what Marines do when the mission fails. Regrouped. Polished the resume. Made calls to every contact from my Raytheon years.

In April 2004, exactly one year after leaving Raytheon, I started at Greentree, a consulting firm specializing in government technology. Yama had taught me lessons I'd carry forward: trust but verify. Read

145

every document as if it could be used against you. Never build something of value without protecting your stake.

THE CLIMB

Management consulting fit me in ways I hadn't expected.

The consulting world spoke a language that resonated with Marine Corps training. Precision. Punctuality. Professional presentation. "If you're early, you're on time. If you're on time, you're late." Could have been lifted from a Marine handbook.

I started on federal contracts with AAFES, the Army and Air Force Exchange Service. Military-adjacent enough that my background gave me an edge. I delivered. Exceeded expectations. Got promoted to manager. Then senior manager.

Each promotion meant more responsibility, more travel, more hours. The family saw less of me with every step up. Kathy managed the house, the kids, the school pickups, the doctor appointments, everything that keeps a household running. I managed projects worth millions of dollars and told myself I was doing it for them. The sacrifice would pay off. The success would justify the absence.

Then the federal business dried up.

AAFES went through restructuring. Budget cuts rippled through our contracts. The pipeline that had fueled my climb narrowed to a trickle.

I refused to let it die.

I found a new vehicle: DBITS, Deliverable-Based IT Services, a contracting mechanism for Texas state agencies. Got on the schedule, learned the procurement process, positioned myself for opportunities others overlooked. Then I won it. Multi-year, multi-million-dollar contract with Texas Health and Human Services Commission. Independent

Verification and Validation of Deloitte's $100M systems integration work.

That contract saved the Texas practice. And it earned me the title of managing director.

There was just one catch. All the work was in Austin. We lived in Dallas. And I was the onsite leader.

TWO CITIES

The first year, I commuted. Every week, same routine. Sunday night or Monday morning, drive three hours to Austin. Check into the Embassy Suites off 183 on Stonelake. Work the contract. Thursday or Friday, head back to Dallas. Weekends with the family, when I wasn't catching up on deliverables.

Exhausting. Unsustainable. But working, professionally at least.

In 2011, I sat down with Kathy.

"We have to move to Austin. I can't do this forever. The contract is multi-year. The kids could start fresh at new schools. We'd actually be together again."

She rejected it. Kids were settled. Her friends were in Dallas. Their friends were in Dallas. "It would uproot the kids" became the final word.

So I got my first Austin apartment that year. A small place in the Domain, functional, close to my clients. The plan was still to commute, still maintain the marriage across two hundred miles of highway.

But something shifted when I signed that lease. I wasn't just working in Austin anymore. I was living there.

Physical separation became emotional separation. Two cities. Two lives. A marriage that existed on paper and weekend visits and phone calls that grew shorter each month.

By 2013, the Dallas house sat half-empty. Kathy and the kids were there. I wasn't. Not in any way that mattered.

I told myself it was work. A marriage could survive distance. Lots of military families did it. We'd done it ourselves during my deployments to Kuwait and Saudi Arabia.

But deployment implies a return date. This had none.

The financial reality was impossible, but I refused to see it. Two households meant nearly double everything. The Dallas mortgage was $360,000, plus a home equity line we'd tapped for $100,000 during the good years. Credit cards bridged the gap between what we earned and what we spent, balances climbing month after month.

I had an MBA. I'd completed Certified Financial Planner course-work. I knew better. I talked to other people about emergency funds, living below your means, the mathematics of compound interest working against you when you're in debt.

And I did none of it myself.

THE LAYOFF

September 2014. The state contract I'd built my practice around didn't renew. Budget cuts, changing priorities, the usual bureaucratic explanations that meant nothing when you were staring at the consequences.

One day I was a Managing Director with six figures. The next I was unemployed.

I didn't allow myself to process it. Not yet.

The very next morning, I formed an LLC. KFidelis Consulting, registered to do business in Texas. Started pursuing SDVOB certification, Service-Disabled Veteran-Owned Business. Networked aggressively. Called every contact I'd made in a decade of consulting. Bid on every relevant opportunity.

The Marine Corps had trained me for this. Adapt and overcome. Don't wallow. Take action.

None of it worked. Not fast enough.

October. November. December. No income. No signed contracts. January. February. The savings I didn't have continued not existing. The credit cards I was already maxed on stayed maxed. The bills kept arriving.

Six months of aggressive networking, proposals, follow-ups. Six months of learning that discipline and effort don't always produce immediate results.

THE DIVORCE

The marriage didn't end with a single conversation. It ended in fragments, over months, across two cities.

But the moment Kathy decided it was truly over came down to Yelp reviews.

During my years of consulting travel, I'd reviewed restaurants obsessively. Austin, Houston, San Antonio, wherever work took me. Client dinners. Solo meals at hotel bars. Business development lunches. I documented all of it, chasing Yelp Elite status like it was another achievement to unlock.

Christian found those reviews.

My son, doing what teenagers do, went looking for information about his increasingly absent father. He found a trail of restaurants in cities that weren't Dallas, meals that weren't with his mother, a life that looked nothing like the one we supposedly shared. He showed Kathy.

To her, it was evidence. I'd been traveling, eating out, living some secret life. The only explanation was betrayal.

She believed I'd been cheating. I hadn't. The marriage had failed for a hundred reasons, but infidelity wasn't one of them. The distance, the finances, the slow erosion over years of prioritizing work over family. Those were real. But there was no affair. No secret relationship. Just a man eating alone in hotel restaurants and reviewing them on the internet like it mattered.

How do you explain that? That sometimes relationships just break? That two people can love each other and still be unable to stay together?

I couldn't. She didn't believe me. By that point, maybe it didn't matter what she believed.

So I gave up and didn't fight for it.

The divorce was filed in May 2015. After twenty-one years, we both knew it was over.

Dividing assets should have been complicated, but there was nothing left to divide. We'd refinanced the house so many times, pulled so much equity, that the sale barely covered closing costs. I agreed to assume all remaining debt. Kathy hadn't worked in years. The obligation was mine.

Final tally: credit card balances exceeding $200,000. Personal loans totaling $75,000. Vehicle debt of $35,000. Monthly obligations over $7,500 with zero reliable income.

By late 2015, I was sitting in a bankruptcy attorney's office, filling out forms that documented every poor decision I'd made for the past decade.

TELLING THE KIDS

The worst part wasn't the bankruptcy. It was what my children carried away from it.

Christian blamed himself. He was the one who found the Yelp reviews. He was the one who showed them to his mother. In his mind, he had set something in motion. Caused the divorce by uncovering evidence of a betrayal that never happened.

I told him the truth. The marriage was already broken long before he went looking. He didn't break it. The reviews weren't evidence of cheating. They were evidence of a father who traveled too much and coped by rating restaurants on the internet. Not exactly a noble defense, but an honest one.

I don't know if he fully believed me. For a long time, he carried the weight of thinking he'd destroyed his parents' marriage. That's the cruelty of divorce. Children always find a way to make it their fault. No matter how many times you tell them otherwise.

Anabel handled it differently. She went quiet. Pulled inward. Processed whatever she was feeling in private. She didn't accuse or question or rage. She simply withdrew.

We've never fully recovered that relationship. She's an adult now, but we don't talk. Not really. The divorce created a wound that hasn't healed. May never heal.

Some consequences don't have remedies.

THE 400-SQUARE-FOOT APARTMENT

The studio was 400 square feet of enforced minimalism.

After the bankruptcy, after the divorce, after everything, I rented the smallest livable space I could find. A complex that had seen better decades, hallways that smelled like industrial cleaner and other people's dinners. Rent was $900 a month, cheap by Austin standards even in 2015.

I owned almost nothing. A mattress on the floor because I hadn't gotten around to buying a frame. A folding table that served as desk, dining surface, and workspace. A single pot, a single pan, two plates, two forks.

The mirror in the bathroom was where I conducted my morning ritual. Every day, I looked at who I'd become and asked the same questions. Who are you without the job title? Who are you without the house in Dallas? Who are you without the lifestyle you used to hide behind?

The answers came slowly.

I was someone who could cook rice and beans without looking at a recipe. Someone who walked to work when gas money got tight. Someone who tracked every dollar in a spreadsheet because the alternative was losing everything again.

The silence was oppressive at first. No family noise, no children, no spouse to fill empty spaces with conversation. Just me and the hum of the air conditioner and distant traffic on I-35.

But the silence became useful. It forced introspection I'd been avoiding for years. All those nights at restaurants with clients, all those trips between Austin and Dallas, all that motion that felt like progress. I'd been running. Escaping the reality of what my life had become.

In 400 square feet, there was nowhere to run.

The Marine Corps taught me that training happens under stress. You can't know what you're capable of until you've been stripped to essentials and forced to perform anyway. Boot camp does this deliberately. Breaks you down to build you up stronger. Bankruptcy did the same thing. Less deliberately. Far more painfully.

What I remember most about that period isn't the paperwork or the court appearances or the calls from creditors. It's the silence. Long stretches alone in that Austin apartment, trying to figure out who I was without the trappings of success I'd used to define myself.

Job titles gone. Marriage gone. Money gone. Relationship with my daughter gone.

What remained was the question I'd been avoiding: Who was I, really, when everything I'd built had been stripped away?

The Marine Corps had an answer. When you're at the bottom, you rebuild. You don't wallow. You assess the situation, identify your resources, start moving forward. One step at a time.

The climbing started there. In that empty apartment. In that bankrupt silence. In that moment of absolute zero.

The only way forward was up.

THE FIRST CONTRACT

April 2015. Six months after the layoff, the phone finally rang with good news.

A three-month database upgrade project for a state agency. The hourly rate was modest, about what I'd made as a junior developer at Raytheon fifteen years earlier. But money wasn't the point. What that contract provided was a chance to rebuild credibility through actual work rather than past titles.

The LLC I'd formed the morning after my layoff. The SDVOB certification I'd pursued through months of paperwork. The network I'd maintained through six months of silence. All of it finally connected.

Not in the dramatic turnaround I'd hoped for. In a small project that paid enough to cover child support and keep the lights on.

One state agency referred me to another. A small project expanded into ongoing maintenance work. By fall 2015, monthly revenue stabilized between $5,000 and $6,000. Far cry from my Managing Director salary, but income I'd earned through actual value creation rather than borrowed money.

The consulting grind wasn't glamorous. Early mornings writing proposals. Late nights finishing deliverables. Weekends on administrative work nobody was paying for.

But for the first time in years, the money coming in was real. Sustainable. Mine.

The 400-square-foot apartment was still small, still sparse. But it no longer felt like punishment. It felt like a base of operations.

I was still alone most of the time. Still eating rice and beans. Still tracking every dollar.

But I was no longer in freefall. I was climbing.

What I didn't know yet was that the next phase of rebuilding wouldn't be something I did alone.

CHAPTER 11

REBUIDING

011200ZOCT15

Who are you when everything you built is gone?

By October 2015, I'd had six months to sit with that question. The bankruptcy was discharged. The divorce was final. The consulting business generated just enough to cover the basics: child support, rent on a 400-square-foot studio, food. I owned a mattress, a small table, and not much else. Forty-one years old, and I could fit my entire life in one room.

The Marines had a term for this: combat ineffective. That's how I felt. Combat ineffective. Still breathing, but unable to do much more than hold the position and wait for reinforcement.

The reinforcement came from directions I never expected. A raid leader from a video game who'd become my business partner and would eventually stand as my best man. A woman I met at a bar who heard my entire disaster of a story on our first date and didn't walk away. A body I finally decided to stop destroying and start rebuilding.

CASEY ROBINSON

The best hire I ever made started in a video game.

I met Casey Robinson in 2006 playing World of Warcraft. For those unfamiliar, WoW is a massively multiplayer online game where players form guilds, coordinate complex raids requiring dozens of people working in precise synchronization, and build communities that can feel as real as any office or neighborhood. It sounds ridiculous when you say it out loud. It was also where I found one of the most important relationships of my professional and personal life.

By 2009, I was the guild master of an organization with over a hundred players, server-ranked and world-ranked in raid progression. Running a competitive WoW guild is like managing a small company staffed entirely by volunteers who can quit anytime they get bored. You learn fast who can lead under pressure, who stays calm when everything goes sideways, who thinks strategically while others panic.

Casey was my raid leader. Night after night, I watched him coordinate forty people through encounters that required split-second timing and zero tolerance for error. He never yelled. Never lost his composure. Never uttered a single bad word. When we wiped, he analyzed what went wrong, adjusted the strategy, and went again. When personalities clashed, he defused the tension without taking sides. When we succeeded, he deflected credit to the team.

I thought to myself: I need this guy in my actual business.

After winning the HHSC contract in 2010, I took a leap of faith on everything I'd observed over four years of gaming together. I called his wife Amanda first, because I'd played with her too and I knew the rule: happy wife, happy life. She gave her blessing.

Then I made Casey an offer as soon as he came online after returning home from work. He just had to move from Inola, Oklahoma to Austin, Texas.

He took it. Packed up his life and relocated right around the same time I moved into my Embassy Suites on Stonelake, beginning the commute that would eventually unravel my first marriage. We were both starting new chapters, neither of us knowing where they would lead.

The work we did together at HHSC went beyond the IV&V contract that had brought us there. Tom Suehs, the Executive Director of the agency, personally asked us to perform an organizational assessment of the entire CIO and IT organization. He felt something was fundamentally wrong with it, something that couldn't be fixed with technology alone.

We spent three months on that assessment. Interviews, process analysis, organizational mapping. The kind of consulting work that requires you to tell powerful people things they don't want to hear. The results led to a major restructuring of the IT organization, including the removal and replacement of key leaders. The CIO was out.

I still have trouble believing I was engaging at that level. Consulting directly to the head of a $40 billion agency, advising the person appointed by the Governor of Texas on how to fix his organization. It was a long way from the supply warehouse at Camp Pendleton.

The restructuring earned us a reputation. Not everyone appreciated our work. Some of the people who remained after the changes saw us as the hatchet men, like the consultants from Office Space who show up to fire people. We had difficult clients within the agency, people who resented what we'd done even if the changes were necessary.

Casey's calm saved us more than once. His demeanor, that same steadiness I'd watched him display leading raids, smoothed over relationships that could have gone sideways. He never took the bait when people got hostile. Never escalated when escalation would have felt justified. He just kept doing good work and treating people with respect until the respect came back around.

When I told him about the bankruptcy, he was genuinely surprised. I'd done such a good job hiding my financial difficulties that even Casey, who worked alongside me every day, hadn't seen it coming. But he understood. He'd been through his own setbacks years earlier. He knew the shame, the way society treats those who've fallen from financial grace.

More importantly, he knew the path back. His perspective became my mantra. Failure isn't fatal unless you let it define you. The qualities that led to my mistakes weren't flaws to be eliminated. They were strengths to be channeled differently.

Casey followed me through every chapter after that. From my consulting business to Honeywell, then to Genpact, and to Amazon Web Services. Ten years after I first watched him lead a raid in a video game, he stood next to me as my best man when I married Lacey.

Having him in that role meant the world to me. I have a tight circle of trust, especially after the betrayals. To have Casey standing there, one of the very few people I trust completely, someone who knew my financial and relationship past and stuck with me anyway, that was everything.

The moment that captured it all came the day after the wedding. Everyone else had left Vegas. Lacey and I and Casey and Amanda had a couples dinner at Hell's Kitchen. Afterward, we sat in the lounge at the Aria where we were staying, drinking Macallan 18 and smoking Cohibas.

We talked about where we'd started. A video game. Pixels on a screen. Two guys who'd never met in person, coordinating raids and talking strategy through headsets. And where we'd gotten to. Business partners. Best friends. Family, really.

Some people think online friendships aren't real. That you can't truly know someone through a screen. I've learned otherwise. The pixels stripped away the pretense. I saw who Casey was under pressure long before I ever shook his hand. And that knowledge, earned through countless virtual battles, translated into one of the most important partnerships of my life.

158

LACEY

The Iron Cactus near Mopac wasn't the kind of place where you expected your life to change.

It was late 2015. My consulting business had finally started generating income. Not much, but enough to stop the bleeding. I'd moved into a small apartment next door, cheap but livable. Friday nights, I'd walk 150 feet to the Iron Cactus for chips and salsa and a margarita I could actually afford. It was the one indulgence I allowed myself.

She was sitting at the bar when I noticed her. Late twenties, dark hair, the kind of confident posture that suggested she wasn't waiting for anyone to buy her a drink. We've argued about what happened next for years. She insists she approached me. I maintain that I made the first move. The truth is probably somewhere in the middle.

"Lacey," she said, extending her hand.

"David."

The conversation was easy. What do you do? Where are you from? Standard Austin small talk. We exchanged numbers before I left.

I asked her to Fleming's for our first date. A steakhouse was probably too expensive for my situation, but I wanted to do this right. We sat in a booth and talked for hours. Something shifted when she asked about my work. Really asked.

"I'm a consultant. Independent. I work with state agencies on technology projects."

"That sounds interesting." She paused. "But that's not the whole story, is it?"

I don't know how she knew. Maybe something in my voice. Maybe she was just perceptive in ways I'd learn to appreciate.

Over the next three hours, I told her everything. The military career. The corporate climb. The consulting success. And then the fall: the layoff, the divorce, the bankruptcy. I told her about the credit card debt and the house I'd lost and the marriage that couldn't survive my own poor choices. I told her about my kids, about Anabel's silence and Christian's confusion. I told her I was rebuilding from zero, that I had almost nothing to offer except honesty about who I was and how I'd gotten there.

Objectively, the worst first date pitch in history. Here's my failure. Here's my shame. Here's everything that should make you walk away.

She didn't walk away.

"Most people never get that kind of financial education," she said. "They just keep making the same mistakes their whole lives."

I didn't know how to respond. In my experience, people didn't react to failure with curiosity. They reacted with judgment, distance, polite excuses to end the conversation.

"Why doesn't that bother you?" I asked.

She shrugged. "Everyone has a past. The question is what you're doing with your future."

We closed down Fleming's that night. Then we met again the following week. And the week after that.

The early months were both amazing and nerve-wracking. I was coming out of a twenty-one-year marriage. The last time I'd actually dated was over two decades earlier. I felt like meeting Lacey was fate, meant to be, but I also couldn't shake the fear. I had flashbacks to Kimberly in the Marine Corps, someone who wanted to try dating an Asian or an older mature man just to see how it was, then moved on when an ex or someone younger came along.

For the first few months, I pushed to go to her place more. Maybe I felt some shame about the studio apartment. She lived in a luxury apartment in South Austin with a roommate, Kristen, who would later be her maid of honor. Eventually Lacey came over to my place.

There was no reaction at all. Like it didn't matter. Like she didn't care.

After her lease ended, she actually moved in with me. That 400-square-foot studio apartment. Mattress on the floor. Folding table. The two of us in that tiny space, and she never complained. We got engaged there in 2017, then moved to Atlanta together for Honeywell.

The decision to stay in that small apartment was so financially smart. She knew we could afford more, but her response was simple: why? Our best times are spent outside the apartment anyway. We used any extra money on experiences, not things. To this day, we favor experiences over large gifts. Trips together. Dinners. Adventures. Not stuff that sits in a house and depreciates.

Our issues, such as they were, surrounded how friends and family would take our age difference. But it turned out nobody really cared. They only cared that we were both happy. And we were both exactly what we needed for each other at exactly the right time.

Lacey's family were the most welcoming, warm, tight-knit people I'd ever experienced. The Midwest comes out in them. Most of her family lives in Indiana, where she's from. We visit often, sometimes nearly monthly. Her mom, as we got to know each other, jokingly but not really jokingly said she never thought Lacey would get married because she'd never met a guy who could take her energy.

That explains a lot about Lacey. Her openness. Her directness. The way she never holds back on how she feels, ever. And her expectation that I do the same.

GETTING HEALTHY

When I was married the first time, I let myself go.

Overweight. Clinically obese, actually. Borderline alcoholic. Chain smoker. At forty, I looked and felt sixty. The Marine Corps fitness and discipline had left me completely. The eighteen-minute three-mile run was about as far from me as a marathon.

After the divorce, I felt I had to redefine myself. Get healthy. Building the business pushed me in that direction, but it was really Lacey who made it happen.

She demanded I quit smoking. Something about kissing an ashtray being disgusting. But I knew it was more than that. She cared about my health. Nowadays she jokes that I have to stay young so I don't die too soon before her. The age gap humor. But underneath it is something real.

I also had to keep up with a twenty-something-year-old. That was motivation enough.

Within the first two years of dating, and up to our wedding, I achieved nearly my Marine Corps weight and physical condition. I may not have been able to run three miles in eighteen minutes anymore, but I was back. I quit smoking. Ate healthy. Stopped drinking so heavily. Started exercising again.

I dropped fifty pounds.

Now we run 5Ks and 10Ks together. My physical fitness matched my financial fitness, and I give Lacey credit for pushing me consistently on both. She never let me settle. Never let me slide back into old patterns. She expected better, and that expectation made me want to deliver.

The Lessons

The years between bankruptcy and recovery taught me things no financial certification ever could.

I learned that high income masks underlying dysfunction. Making good money can hide destructive patterns until it's too late. Despite earning six figures for years, the inability to handle a modest roof repair without borrowing revealed how unsustainable my financial structure really was. Income isn't wealth. Spending habits determine wealth.

I learned that debt accumulation has mathematical endpoints. There's a point where even high earnings can't overcome accumulated obligations. Credit cards feel like solutions until they become prisons.

I learned that recovery requires support. Mentors like Casey and partners like Lacey provide crucial encouragement during rebuilding. No financial recovery happens in isolation. The people who believe in you when you can't believe in yourself, they're the ones who make rebuilding possible.

I learned that physical health and financial health are connected. When I let my body go, I was letting everything go. Getting fit again was part of getting my life back.

And perhaps most importantly, I learned that past failures could become future strengths. The painful lessons from financial collapse gave me perspective for building more sustainable success. I knew what rock bottom looked like. I never wanted to see it again.

The rebuilding wasn't just about money. It was about becoming someone worth rebuilding into. Someone healthy. Someone honest. Someone capable of real partnership instead of just cohabitation.

Lacey saw that person before I did. Casey believed in that person when I had my doubts. Together, they helped me become him.

CHAPTER 12

FULL CIRCLE

010800ZAPR17

The years after bankruptcy taught me how to survive. What came next taught me how to build something worth keeping. From 2017 forward, everything changed. Not because the world got easier, but because I finally understood how to move through it. The corporate climb resumed, but this time with different priorities. The partnership with Lacey deepened into something I hadn't known marriage could be. And somewhere along the way, I stopped rebuilding and started building. The difference matters more than I can explain.

This chapter spans nearly a decade, from the Honeywell offer that pulled me back into corporate leadership to the Nashville home where I'm writing these words. It includes the career progression I once thought was lost forever—Director, then Senior Director, then Amazon Web Services. It includes a wedding in Vegas, a vow renewal seven years later, and a marriage built on principles my first one never had. It includes the slow, painful work of rebuilding a relationship with my son, and the ongoing silence from my daughter that I've learned to hold without resolution.

The title of this chapter is "Full Circle," but that's not quite accurate. Circles imply returning to where you started. I haven't returned anywhere. The kid from Riverside, the boot camp recruit, the supply Marine in Kuwait—he's still in here somewhere, but he wouldn't recognize the life I'm living now. What came full circle wasn't geography or circumstance. It was something harder to name. A sense of completion that has nothing to do with finishing. The understanding that you can lose everything, rebuild from nothing, and end up somewhere better than where you began—not despite the losses, but because of what they taught you.

HONEYWELL

The Honeywell opportunity emerged from an unexpected source. They were seeking a Director and Deputy CTO to lead their transformation from traditional waterfall development to agile practices.

I dismissed it immediately. The position seemed beyond my reach. I'd been out of corporate leadership for two years, my resume now featured a gap that included bankruptcy, and my agile knowledge was theoretical at best. The self-doubt that had been my constant companion since the fall whispered its familiar refrain: *You're not ready. You're not qualified. You're not the person you used to be.*

That evening, I shared my hesitation with Lacey over dinner at our apartment. "I've never formally led an agile team," I explained. "My experience is all traditional project management."

She didn't accept my excuses. "You understand systems, people, and delivery," she said. "The specific methodology is just a framework you can learn. You've been adapting to new environments your entire career."

When I still hesitated, she pushed harder. "The job description is asking for what you can become, not just what you've already done. Study and get certified in agile this weekend, then talk to them on Monday as someone who's already closing the gap."

Her perspective, unbound by my past failures and limitations, helped me see possibilities I had overlooked. That weekend, I spent sixteen hours completing an accelerated agile certification course. Scrum Master and Product Owner certifications back-to-back. By Monday morning, I wasn't just a candidate for the position. I was actively demonstrating the growth mindset the role required.

The interview process revealed how much my approach to corporate opportunity had evolved. Where I once would have focused on title, compensation, and status, my questions now centered on team dynamics, organizational challenges, and real opportunities for impact. The technical work I'd done during my consulting years paid off. I could speak both languages: the strategic vision of leadership and the practical reality of implementation.

After three rounds of interviews, Honeywell extended an offer. Director and Deputy CTO position with a base salary of $200,000. It represented both stability and opportunity. More importantly, it validated something I'd been afraid to believe: that the skills I'd built, even through failure, had value.

I called Lacey from the parking lot. "They offered it," I said.

"Of course they did," she replied, as if there had never been any doubt.

THE QUIET YEARS

The period from 2017 to 2021 was the first stretch of my adult life without crisis.

No bankruptcy. No divorce. No restaurant collapsing. No layoff. Just steady, intentional progress, the kind of stability I'd never experienced and didn't entirely trust at first.

Honeywell hired me in 2017 at $200,000 a year, more than I'd made as a Managing Director before everything fell apart. The role felt like validation, proof that the years of consulting grind had rebuilt not just my income but my professional reputation. Lacey pushed me to negotiate, to counteroffer, to stop undervaluing myself the way I had when I was desperate. She was right.

We married in August 2018. Vegas, twenty-five guests, no pretense. Casey stood beside me as best man, the same guy who'd been my raid leader in World of Warcraft and my business partner through bankruptcy. Lacey's parents flew in. It was small and intentional, nothing like the first marriage, and that was exactly the point.

From Honeywell, I moved to Genpact to build and lead their global cloud advisory practice. Each transition built on the last. The skills I'd developed through failure, the ability to navigate ambiguity, to translate between technical and business worlds, to lead through uncertainty, they weren't just survival mechanisms anymore. They were differentiators.

By 2020, I was no longer rebuilding. I was building. Emergency fund: funded. Retirement contributions: maximized. Lifestyle: intentionally below our means. The lessons from the 400-square-foot apartment weren't just memories. They were daily practices embedded into how Lacey and I managed our life together.

I didn't know what was coming next. But for the first time, I was ready for whatever it was.

BUILDING A PARTNERSHIP

My relationship with Lacey is night and day from my first marriage.

The differences started with money but went far deeper. With Kathy, I had been the sole provider for twenty-one years. Every dollar came from my paycheck. Every financial decision was mine alone. Every pressure was mine alone. When things went wrong, I had no partner to share the burden or challenge my blind spots.

Lacey was a working professional with her own career, her own income, her own financial identity. At times during our relationship, her earnings matched or exceeded mine. After my bankruptcy, when I was building the consulting business from scratch, her income was significantly higher than mine. She carried us through those lean months without resentment, without keeping score.

That dynamic required something I'd never built before: a system for managing money as true equals.

We developed a proportional split. Rather than dividing expenses fifty-fifty, which would have been impossible when our incomes were drastically different, we contributed to shared expenses based on what each of us earned. If she made sixty percent of our combined income, she covered sixty percent of the rent, utilities, groceries. The system required transparency my first marriage never had. We had to know what each other earned, discuss our individual and shared expenses, negotiate and adjust as circumstances changed.

The proportional system served us through every phase: the lean consulting years, the Honeywell salary jump, Genpact, and eventually AWS. Today, our combined annual income, W-2 earnings plus rental properties, investments, and VA benefits, exceeds one million dollars. The same framework that managed scarcity now manages abundance. Our incomes shifted dramatically over time, but the framework adapted. Neither of us ever felt like we were carrying the other or being carried.

But the partnership extended beyond money.

From the beginning, one of Lacey's most common phrases was "tell me a story." It wasn't a request for entertainment. It was a push to communicate, to open up, to share what I was actually feeling. My default mode had always been to process internally, to keep struggles private, to project competence even when I was drowning. That approach had contributed to the slow collapse of my first marriage.

Lacey wouldn't accept it. She pushed consistently against my communication boundaries until opening up became habit rather than effort. Nothing stays hidden anymore. The good, the bad, the uncertain, we talk through all of it.

We set written goals, personal, career, and financial, both individually and as a couple. Monthly check-ins. Annual reviews. The same deliberate planning I describe in my financial recovery book, The Real Money Guide, we practice in our own lives. I push her to negotiate and counteroffer during salary discussions. She pushes me to adopt more agile and people-centric practices in my leadership roles. We make each other better.

After our wedding in Vegas, we established a tradition: "monthiversaries" on the 25th of every month. We dress up, go out, celebrate our partnership. It sounds small, but it's intentional. We prioritize each other deliberately rather than assuming the relationship will maintain itself.

For our seventh anniversary in 2025, we returned to Vegas for a vow renewal, inviting back much of the same twenty-five people from the original wedding. Standing there, saying those words again, I understood something I hadn't fully grasped the first time around. Marriage isn't a destination. It's a practice. And the practice only works if both people show up intentionally, consistently, every single day.

She saw potential when I could only see wreckage. She saw how we could build something together that neither of us could build alone. What we've created since proves she was right.

AWS

The call came in early 2021, during an impromptu beach trip with Lacey to South Padre Island. We had just checked in to the AirBNB.

"Amazon Web Services," the recruiter said. "Two hiring managers are interested in your background."

I almost laughed. AWS, the cloud computing giant that had transformed the technology landscape. The company that had helped build the infrastructure for half the internet. And they wanted to talk to a guy who'd filed for bankruptcy five years earlier.

But I'd learned something in the years since that Austin apartment: imposter syndrome is just another form of self-sabotage. The skills I'd built, through military logistics, through software engineering, through consulting, through failure and recovery, were exactly what they were looking for. I could translate between worlds, bridge the gap between technical implementation and business strategy. I could lead teams through uncertainty because I'd led myself through worse.

The Loop

The interview process was the most intense I'd ever experienced. Several preliminary rounds led to what Amazon internally calls "the loop," a full day of five hour-long interviews plus a lunch with a "buddy" for candid conversation. Each interview focused on two specific Amazon Leadership Principles such as Customer Obsession, Think Big, Bias for Action, Learn and Be Curious. The interviewers weren't just asking about my experience. They were probing for specific examples that demonstrated how I embodied each principle.

I spent weeks preparing. I mapped every significant experience from my career, the barcode system in Kuwait, the satellite logistics in Korea, the Yama disaster, the bankruptcy recovery, the consulting grind, to Amazon's sixteen Leadership Principles. Each story needed a clear situation, task, action, and result. The process forced me to articulate lessons I'd learned instinctively but never verbalized.

The preparation paid off. When an interviewer asked about a time I'd taken a calculated risk, I didn't fumble for an answer. I told them about developing that supply tracking system with no guarantee it would work, about testing it in the field during Operation Native Fury. When they asked about learning from failure, I had Yama ready. When they asked about customer obsession, I described building systems that Marines in the field actually needed rather than what headquarters thought they wanted.

At every stage, I expected them to see through me, to recognize the bankruptcy on my credit report, the divorce in my past, the years of wandering between Austin and Dallas. They never did. Or maybe they did, and they understood that failure isn't a disqualification. It's a credential.

The offer came on a Tuesday afternoon. The total compensation was nearly ten times what Raytheon had paid me two decades earlier. It was more than twenty times what I'd earned as a sergeant in the United States Marine Corps.

I called Lacey from the same parking lot where I'd called her about Honeywell three years before.

"They offered it," I said.

She didn't miss a beat. **"Of course they did."**

Finding My Place

My first ninety days at AWS consisted of structured onboarding, role-specific training, and deliberate networking. This was unlike any company I'd joined before, small or large. Most places give you hours, maybe days, before expecting you to deliver. Amazon invested in ensuring I understood the culture, the systems, the expectations.

The imposter syndrome hit hard. For at least the first year, it never fully went away. Throughout my career, in every meeting, I'd felt like one of the smartest people in the room. At Amazon, I never felt that way. The caliber of talent was humbling. Every conversation challenged me. Every project pushed my capabilities.

This connects to Amazon's "bar raiser" concept in hiring. The goal is to hire people who are better than fifty percent of the existing Amazonians at that level. The bar keeps rising. You have to embody "Learn and Be Curious" just to maintain your standing, let alone advance.

The role itself fit my background perfectly. I lead an organization of Customer Solutions Managers, similar to Technical Delivery Managers, who support primarily Public Sector clients. Government and government-adjacent projects. The work requires exactly the combination of technical knowledge, process discipline, and stakeholder management that the Marine Corps and consulting had drilled into me.

My job, as I tell my team, mirrors my philosophy from the Marine Corps: troop welfare. The two Marine Corps Leadership Objectives are Mission Accomplishment and Troop Welfare. At Amazon, I map those to Customer Obsession and being Earth's Best Employer. My mission is ensuring my team members have everything they need to deliver for customers, that they're compensated fairly, and that they maintain balance so their personal lives aren't consumed by work.

Some team members joke that the only time I "yell" at anyone is when they respond to a Slack message or email while on vacation. That's not entirely a joke. I've seen too many careers, including my own earlier ones, derailed by failing to protect personal time. The work will always expand to fill available space. Leadership means creating boundaries that others can't or won't create for themselves.

NASHVILLE

We stopped running and started choosing. But the path to Nashville wasn't direct.

After selling our Austin house, we started building a new construction home in Plano, a Dallas suburb. We moved into a temporary apartment at Legacy West, our belongings in storage, watching our future home take shape across the street. We watched every step of construction, just like we did with the Austin home. Everything was on track.

Two weeks before closing. The appraisal came in.

We were in the final stages, needing only a current appraisal before closing. The number came back twenty percent lower than the build price. The builder pushed back, tried to negotiate, suggested we make up the difference in cash. The back-and-forth turned ugly. They threatened mediation or losing all of our $100,000 deposit.

I activated the VA Escape Clause in our contract. A little known (to the builder) provision that allows veterans to withdraw without penalty on VA Home Loans if the appraisal doesn't support the purchase price. The builder fiercely resisted. They pushed for arbitration again, tried to hold us to the contract anyway. I retained an attorney to send legal notices on contract performance. The VA appraisal clause was rock solid. They had no case and eventually they knew it. But convincing them took weeks of pressure.

All of this unfolded while we were visiting Lacey's family in Indiana. Suddenly, we had no future home, over $100,000 in deposits tied up in legal limbo, and all our possessions in a Dallas storage unit.

On Thursday, November 10, 2022, the Marine Corps birthday, we were officially released from the contract with the builder. Ooh-Rah, what a birthday present! Our deposits were returned. We were free. And homeless.

Nashville had been one of three cities we'd considered moving to after Austin, along with Dallas and Charleston. That evening, sitting in

Lacey's mother Rella's living room, we made the decision. Nashville it was.

I reached out to Chris Mannino, a realtor referred by a coworker who lived in the area. I gave him our list of wants and needs. By the end of the day, he'd sent back roughly a hundred listings that matched our criteria. Friday morning, Lacey and I narrowed it to five homes we wanted to see. By Friday evening, Chris had all five scheduled for Saturday viewings.

We packed our bags, drove straight down I-65, and got a room at the Aloft on I-65 in Franklin, Tennessee. Saturday morning at 8 AM, we started looking at houses.

By noon, we were at the fourth home. Lacey and I looked at each other and knew. This was it. A house on a hill in Brentwood, neighbors who would wave from their driveways, a community that felt like community.

The asking price had just been reduced by $200,000 the day before. The seller had paid $600,000 more than that just six months earlier. We wrote the offer at full price and sent it off, then started the ten-hour drive back to Dallas. Three hours into the drive, before we even got to the state border, Chris called. Our offer was accepted with no conditions.

We closed on December 7, 2022, less than a month after the Dallas contract collapsed. From homeless to homeowners in twenty-seven days.

The irony wasn't lost on me. Twenty-five years earlier, I'd enlisted in the Marine Corps to escape the suburbs, to find something bigger than the life my parents had mapped out for me. Now I was choosing a version of that life: the house, the yard, the settled existence. But I was choosing it. That made all the difference.

THE CHILDREN

Some wounds don't heal on schedule.

Anabel is an adult now, finding her own path. After the divorce, she started in nursing, then shifted to gymnastics coaching. I don't know where her career has taken her since. We don't talk. Not really.

The silence didn't come all at once. After the divorce in 2015, our communication tapered slowly. I think she believed I was the cause of her mother's pain and financial struggles. I think she suspected outside influences, infidelity perhaps, though that wasn't the reality. The truth of why marriages fail is always more complicated than any single narrative, but complicated truths don't satisfy the need for clear villains.

The final break came in a way I didn't anticipate. When Lacey and I married in August 2018, we kept the wedding small. Twenty-five guests, adults only, a Vegas ceremony that felt right for where we were in our lives. I didn't invite Anabel or Christian. I told myself I was protecting them, not wanting to expose too much about my new relationship too quickly. In hindsight, that was probably a mistake.

In 2020, both kids visited us in Austin. That's when I told them Lacey and I were married, that we'd been married for two years. I could see it hit Anabel differently than Christian. The discovery that I'd hidden something that significant, even with good intentions, felt like another betrayal. We texted a few times after that visit, but her responses grew shorter. After May 2021, she stopped replying altogether.

I've attempted to reconnect since. Each attempt has met silence. I've learned to take a passive approach now, leaving the door open, making clear I'm available, but allowing her to set the timeline for when, or if, she's ready to engage.

It brings me pain. I won't pretend otherwise. The relationship I have with Christian makes the absence of connection with Anabel more acute, not less. I hope we can resolve whatever she believes happened, that we can reengage before too many more years pass. But I can't force it. And regret is just another form of self-indulgence. What matters is being available if she ever decides to reach out.

Christian took a different path.

We talk and text consistently now. I helped him think through college decisions, evaluate internships, and ultimately choose Meta for his first role after graduation. Financial advice flows freely. How to handle RSUs, the importance of 401(k) matching, the same lessons I wish someone had taught me at his age.

Lacey and I attended his graduation from Texas A&M on May 14, 2022. Watching him walk across that stage, knowing the relationship we'd rebuilt from the wreckage of those early post-divorce years, I felt something I hadn't expected: hope. Not just for him, but for what's possible when you stay present and available even when presence feels impossible. Two months later, he started at Meta, launching his career in tech just as I had launched mine at Raytheon two decades earlier.

Christian's girlfriend Michelle, now his fiancée, has built a genuine relationship with Lacey. Family dinners feel like family dinners. The future has the shape of something I can recognize and embrace.

Some circles do close. Not all of them, not yet. But enough to believe that time and presence can heal what seemed irreparable.

STILL BUILDING

I'm fifty-one years old.

I lead an organization at one of the most innovative companies on the planet. I own a home in Tennessee with a woman who saw potential when I could only see wreckage. I have a relationship with my son that was rebuilt from the debris of divorce. I have a career that somehow absorbed every failure and converted it into qualification.

None of this was inevitable. There were moments, more than I care to count, when the story could have ended differently. The bankruptcy could have become permanent defeat. The divorce could have severed me from my children entirely. The layoff could have been the final word on my professional life.

It wasn't. Not because I'm exceptional, but because I kept moving. Because the Marine Corps taught me that the only failure that matters is the failure to get back up. Because people like Casey and Lacey believed in me when belief felt irrational.

The kid who rebuilt carburetors in Mr. Redwine's shop, who built supply systems in Kuwait, who built and lost a restaurant, who rebuilt himself from bankruptcy, he's still building. That's what Marines do. That's what I do.

I don't know what comes next. Another role, another challenge, another opportunity to fail or succeed. But I know how I'll approach it: with discipline earned through disaster, with humility learned the hard way, with partnership that makes the work worthwhile.

The mission continues. It always does.

EPILOGUE

151200ZDEC25

If I could talk to that kid in Riverside, the one who felt trapped by his parents' expectations, who stole from stores to feel something, who saw the Marines as escape rather than opportunity, I'm not sure what I'd tell him.

Maybe this:

> *The path you're about to take will be harder than you imagine and more rewarding than you can conceive. You'll succeed beyond your wildest dreams, and you'll fail in ways that feel like the end of everything. Neither will define you. What will define you is what you do next, always what you do next.*

The military service you're about to begin will reshape you in ways both visible and invisible. The uniform will come off eventually, but the transformation will be permanent. You'll learn that bullets don't fly for supply Marines, but that doesn't mean the mission doesn't matter. Every system you build, every problem you solve, every team you support will be part of something larger than yourself.

The failures, betrayals, and disappointments will come. The restaurant you'll pour yourself into will be taken from you. The marriage you'll build over twenty-one years will crumble under the weight of choices you haven't yet made. The money you'll earn will disappear through a combination of bad luck and worse decisions. You'll sit in a bankruptcy attorney's office wondering how you got there.

But you'll get up. You'll always get up. That's what Marines do, even after the uniform is gone.

And on the other side of all of it, on the other side of the falling and the getting up, the building and the losing, the mistakes and the lessons, you'll find something you didn't expect. Peace. Not perfection, but peace. A partner who sees you clearly and loves you anyway. Work that challenges and rewards. Enough money to stop worrying about money. And the knowledge that every hard thing you survived made the good things possible.

The composition is still being written. Every day adds another note, another unexpected harmony. But if there's one thing I've learned, it's this: the music is in the making, not the finishing. And the most beautiful passages often come after the darkest ones.

FINAL SALUTE

To my fellow veterans, especially those who served in support roles: your service mattered. The bullets you tracked, the supplies you managed, the systems you built, they were the invisible architecture that made everything else possible. Don't let anyone tell you that non-combat service is lesser service. The military is a machine with countless moving parts, and every part is essential.

To anyone rebuilding from failure: it's possible. Not easy, not quick, but possible. The skills you've developed, the lessons you've learned, the scars you carry, they're not liabilities. They're credentials. The world needs people who've been through the fire and come out the other side. Your story isn't over. It's just entering a new movement.

And to whoever's reading this: the road is longer than it looks, harder than it seems, and more beautiful than you can imagine. Keep walking.

Semper Fidelis.

David Kim

In Memoriam

Cha Yeon Kim (1903–2000)

Before Mr. Redwine's auto shop, before the Marine Corps, before any of the mentors and partners who shaped my adult life, there was my maternal grandmother. She survived Japanese occupation, emigrated to America with nothing, and spent her final years sharing a bedroom with a grandson who was too young and too angry to appreciate what he had.

She broke house rules to bring me food when I was being punished, pulling a hoagie from her pants, crying, calling me an asshole in Korean. It was the first time I understood that love could be bigger than rules, that someone could risk something for you simply because you mattered to them.

I didn't get to thank her before she died in 2000. But her fingerprints are on every page of this book. The resilience I write about, the getting back up, the refusing to quit, I learned it from her before I learned it from anyone else.

EARNEST PAUL REDWINE (2/14/1943-5/6/2021)

In a small auto shop at Riverside Poly High School, Mr. Redwine created more than a space for learning mechanical skills. He built a sanctuary where young minds could discover their potential.

A US Navy Sailor who became an educator, Mr. Redwine understood that true teaching goes beyond technical instruction. His approach to mentorship was as methodical as the engine rebuilds he supervised, yet as nurturing as his quiet encouragement.

His legacy lives on not just in the students he taught, but in the lives they touched in turn. In his passing, we lost not just a teacher, but an architect of character who understood that every young person, like every car, holds untold potential waiting to be revealed.

LORETTA CORNETT-HUFF (6/16/1932-6/20/2013)

Loretta Cornett-Huff was more than an educator; she was an innovator who saw possibilities where others saw obstacles. In her 42 years of government service, she changed the face of military education, but her true gift was in changing lives.

Her office at the Kaneohe Base Education Center was more than an administrative space; it was an incubator of dreams where thousands of service members found paths to education that they never thought possible.

Her legacy lives on in the thousands of military members whose lives she touched, in the educational programs she championed, and in the success stories she helped write. In her passing, we lost not just an educator, but a visionary who understood that education is the most powerful force for transformation.

PHOTOS

My Grandma

My Aunt

Boot Camp Family Day 1992

March Air Force Base Tarmac 1994

MCRD San Diego 1992 (3rd row from top, 3rd from left)

Shannon Airport, Ireland 1994

David and Leonard Hill Korea 1998

Glenn and Anabel, Hawaii 1998

David and Casey Las Vegas 2018

187

Our Wedding, Las Vegas 2018

Christian's Graduation 2022   Vegas Vow Renewal 2025

ACKNOWLEDGMENTS

Writing a memoir is an act of excavation. You dig deep through years of experience, unearthing memories you'd forgotten and confronting ones you'd tried to bury. No one does that digging alone.

My wife, Lacey Caplinger. You came into my life when I had nothing left. Where others might have seen a man defined by his failures, the bankruptcy, the divorce, the restaurant that was stolen, you saw someone worth betting on. You saw potential when I could only see wreckage. Every page of the chapters after we met is really our story, not mine. Thank you for choosing to write it with me.

My son, Christian. Our relationship has been complicated by distance, divorce, and years I can't get back. You've seen me rebuild. You've watched me make different choices. I hope someday these pages help explain the father I was trying to be, even when I fell short. I love you more than I've ever been able to say properly.

My daughter, Anabel. The silence between us is one of the hardest things I carry. I don't know if you'll ever read this, but if you do, I want you to know: the door has never closed. I've made my mistakes. I've owned them in these pages as honestly as I know how. None of them changed how much I love you. I hope someday we find our way back to each other. Until then, you remain part of my story and I hope I remain part of yours.

My grandmother, Cha Yeon Kim. You died in 2000, but your influence echoes through this entire book. That smuggled hoagie in our shared bedroom, pulling it from your pants, tears streaming, calling me an asshole in Korean, taught me more about love, rules, and doing what's right than any formal education ever could. You survived the Japanese occupation, crossed an ocean, built a life in a foreign country, and still found ways to give. When I write about resilience, I'm writing about you. I wish you could have seen this book.

My aunt, Sun Lye Lim. That Commodore 64 you bought me changed everything. You saw something in a troubled kid that his own parents couldn't see, and you invested in it. The computer that seemed like just a toy became the foundation of my entire career. Thank you for seeing what I could become before I could see it myself.

My mother and step-father, Hwa Ja Yoon and Kil Young Yoon. Our relationship has been complicated, and this memoir doesn't hide that. The expectations you had, the cultural pressures you carried, the discipline you administered, I've spent decades processing all of it. But you also sacrificed to give me opportunities. You worked jobs you didn't love so I could have choices you never had. The kid who ran away from home, who joined the Marines partly to escape your rules, grew into a man who understands, at least partially, why you did what you did. We're still working on the rest.

Mr. Earnest Redwine. You died before I could tell you what you meant to me, and that failure haunts me still. In your auto shop, covered in grease and engine oil, you taught me that discipline wasn't about following rules; it was about caring enough to do something right. You saw a troubled kid and decided to invest in him anyway. I've tried to pay that forward. I hope I've succeeded.

LtCol Kathleen Murney. You were the first officer to treat me like I had potential beyond my rank. In the SASSY Management Unit, you gave me room to innovate, to fail, to try again. The barcode system that earned me recognition in Kuwait and Saudi Arabia was only possible because you created space for a junior Marine to think beyond his job description. You taught me that leadership wasn't about giving orders; it was about developing the people around you. I've tried to carry that lesson into every leadership role I've held since.

Sergeant Mike Sweetsir. You treated a boot Lance Corporal like someone worth mentoring. You didn't have to. NCOs had a thousand other demands on their time. But you chose to invest in a kid who was still figuring out how to be a Marine. That investment paid dividends you'll never fully see.

Staff Sergeant Leonard Hill. Our partnership during Operation Foal Eagle was one of the highlights of my military career. Those cold nights in Pohang, troubleshooting connections and debugging our supply systems, taught me what true collaboration looks like. You trusted my technical skills. You respected my cultural background. You treated me as a partner, not just a subordinate.

Glenn Knepp. Thirty years and counting. From that locked balcony door in Hawaii to our constant new ventures, you've been a constant. Innovative Creations taught me that I could build something outside the structure of military service. Every late-night discussion about business models, certifications, jamming out on our guitars, and enjoying Korean food, every technical problem we solved together, every beer we shared after a long day; it all added up to a partnership that has outlasted marriages, careers, and geographic separations. Two guys who met in the Marines, still convinced we can build something better. Here's to the next thirty.

Jimmy Parker. Those smoke breaks during class at Hawaii Pacific University gave me more practical business education than any textbook. You saw that I wasn't just learning to code; I was building a toolkit for a life you knew I'd have to build. The introductions to Ingram Micro and Tech Data, the business planning guidance, the push to think like an entrepreneur instead of just a technician; you prepared me for a civilian world that the Marine Corps never could. Thank you for seeing beyond the classroom.

Loretta Cornett-Huff. At the base education center, you helped so many Marines find their way through the complicated path from military service to civilian education. You made me your "poster child," and together we showed other service members what was possible. Your patience with paperwork, your encyclopedic knowledge of educational benefits, and your genuine care for every Marine who walked through your door made a difference in more lives than you'll ever know.

My fellow Marines. This book changes names and details to protect privacy, but the experiences are real. To everyone who served alongside me, in boot camp, at Camp Pendleton, in Kuwait, Saudi Arabia, Hawaii, and Korea, thank you. You taught me what it means to be part of something larger than yourself. The bonds we formed aren't diminished by time or distance. Semper Fidelis.

And finally, **to anyone rebuilding from failure.** I've been where you are. The bankruptcy, the divorce, the moment when everything you built collapses and you can't see a way forward. It's possible to get up. It's possible to build something better. The skills you've developed, the lessons you've learned, the scars you carry, they're not liabilities, They're credentials. Keep going.

GLOSSARY

This glossary provides definitions for military terms, acronyms, and select financial concepts used throughout the book.

1st FSSG: 1st Force Service Support Group. A major logistics command at Camp Pendleton responsible for combat service support.

96: A 96-hour liberty pass, giving Marines four days off duty.

AAFES: Army Air Force Exchange Service. The retail and services organization for military installations.

Barracks: On-base housing for unmarried enlisted service members.

BAS: Basic Allowance for Subsistence. Monthly food allowance for service members.

BAH: Basic Allowance for Housing. Monthly housing allowance for service members not living in barracks.

Boot Camp: Marine Corps Recruit Training. The initial 13-week training program that transforms civilians into Marines.

Chow Hall: Military dining facility.

Conex: Container Express. A standard military shipping container used for transporting and storing equipment.

CSSG-3: Combat Service Support Group 3. A logistics unit based at Marine Corps Base Hawaii.

DD-214: Certificate of Release or Discharge from Active Duty. The official document issued upon separation from military service.

Deployment: A period during which service members are sent to a specific location to perform military duties, typically overseas.

Dependents: Family members (spouse, children) of a service member who are entitled to military benefits.

DI: Drill Instructor. The Marine Corps NCOs responsible for training recruits at boot camp.

Dress Blues: The formal blue uniform worn by Marines for ceremonial occasions.

E-Club: Enlisted Club. A recreational facility on military bases for enlisted personnel.

EAS: End of Active Service. The date when a service member's enlistment contract ends.

Field Day: Mandatory intensive cleaning of barracks and workspaces, typically conducted weekly.

GP Tent: General Purpose tent. Standard military field shelter.

Honorable Discharge: The most favorable type of military discharge, indicating satisfactory service.

Liberty: Authorized free time away from duty, similar to civilian time off.

MCAS: Marine Corps Air Station. An air base operated by the Marine Corps.

MCB: Marine Corps Base. A Marine Corps installation.

MCT: Marine Combat Training. Basic infantry training for non-infantry Marines.

MCRD San Diego: Marine Corps Recruit Depot, San Diego. One of two locations where Marine Corps recruit training is conducted.

MOS: Military Occupational Specialty. A code identifying a specific job within the military.

MPS: Maritime Prepositioned Ships. Vessels loaded with military equipment and supplies positioned in strategic locations worldwide.

MRE: Meal, Ready-to-Eat. Individual field rations in lightweight packaging.

NCO: Non-Commissioned Officer. An enlisted service member holding a rank of corporal or sergeant.

Oorah: A Marine Corps battle cry expressing enthusiasm, motivation, or acknowledgment.

PCS: Permanent Change of Station. A relocation to a new duty station, typically for an extended period.

POG: Person Other than Grunt. Slang (sometimes derogatory) for non-infantry Marines.

PT: Physical Training. Organized exercise and fitness activities.

PX: Post Exchange. A retail store on military installations (Army term; Marines typically use "Exchange").

Rack: Military slang for bed or bunk.

Reenlistment: The act of signing a new contract to continue military service after the initial enlistment ends.

SASSY: Supported Activities Supply System. The Marine Corps' primary automated supply system.

Semper Fi: Short for "Semper Fidelis" (Latin for "Always Faithful"), the Marine Corps motto.

SMU: SASSY Management Unit. The organizational unit responsible for operating the SASSY system.

SNCO: Staff Non-Commissioned Officer. Enlisted Marines in ranks E-6 through E-9.

SOP: Standard Operating Procedure. Established protocols for routine operations.

TAD: Temporary Additional Duty. A short-term assignment away from one's primary duty station.

Terminal Leave: Accumulated leave taken at the end of military service before official discharge.

OPERATIONS REFERENCED

Operation Native Fury (March–April 1994): Combined exercise in Kuwait involving U.S. and coalition forces.

Operation Vigilant Warrior (October–November 1994): U.S. military response to Iraqi troop movements toward Kuwait.

Operation Foal Eagle (October–November 1998): Annual combined exercise between U.S. and Republic of Korea forces.

VETERAN RESOURCES

Crisis Support
- **Veterans Crisis Line**: 988 (Press 1) or text 838255
- **Crisis Text Line**: Text VETERAN to 741741
- **Suicide Prevention Lifeline**: 988

VA Benefits & Healthcare
- **VA.gov**: Official portal for all veteran benefits
- **1-800-827-1000**: VA Benefits Hotline
- **MakeTheConnection.net**: Mental health resources
- **MyHealtheVet** (MyHealth.VA.gov): Online health management portal

Career Transition
- **American Corporate Partners** (ACP-USA.org): Veteran mentorship program
- **LinkedIn for Veterans**: Free Premium access for transitioning service members
- **Hiring Our Heroes** (HiringOurHeroes.org): U.S. Chamber of Commerce veteran employment program
- **RecruitMilitary** (RecruitMilitary.com): Job board for veterans

Education Benefits
- **GI Bill Comparison Tool** (VA.gov/gi-bill-comparison-tool)
- **CLEP and DSST Testing**: Earn college credit for military training
- **Pat Tillman Foundation** (PatTillmanFoundation.org): Scholarships for veterans and military spouses
- **Student Veterans of America** (StudentVeterans.org): Campus chapters and advocacy

Financial Assistance & Literacy
- **Military OneSource** (MilitaryOneSource.mil): Financial counseling and resources

- **FINRA Foundation Military Financial Readiness**: Free financial education
- **Operation Homefront** (OperationHomefront.org): Emergency financial assistance

Housing & Homelessness

- **SSVF** (VA.gov/homeless/SSVF): Supportive Services for Veteran Families
- **National Call Center for Homeless Veterans**: 1-877-424-3838
- **Habitat for Humanity Veterans Build** (Habitat.org): Homeownership programs

Family Support

- **Blue Star Families** (BlueStarFam.org): Military family support network
- **National Military Family Association** (MilitaryFamily.org): Programs and scholarships
- **TAPS** (TAPS.org): Tragedy Assistance Program for Survivors

Health & Wellness

- **Give an Hour** (GiveAnHour.org): Free mental health services
- **Cohen Veterans Network** (CohenVeteransNetwork.org): Mental health clinics for veterans and families
- **Wounded Warrior Project** (WoundedWarriorProject.org): Programs for wounded veterans
- **Team Red White & Blue** (TeamRWB.org): Physical and social activity programs
- **The Mission Continues** (MissionContinues.org): Community service fellowships

Disability Claims & Appeals

- **DAV** (DAV.org): Free claims assistance
- **Veterans of Foreign Wars** (VFW.org): Accredited claims representatives
- **American Legion** (Legion.org): Veterans service officers
- **AMVETS** (AMVETS.org): Claims assistance and advocacy

HISTORICAL NOTE

The events described in this memoir took place between 1991 and 2025. While the author has made every effort to accurately represent his experiences and the historical context in which they occurred, some names, dates, and identifying details have been changed to protect the privacy of individuals involved. Dialogue has been reconstructed from memory and may not represent exact quotations.

ABOUT THE AUTHOR

David Kim served in the United States Marine Corps from 1991 to 1999 as a supply administration specialist. His deployments included Kuwait (Operation Native Fury), Saudi Arabia (Operation Vigilant Warrior), and South Korea (Operation Foal Eagle), where he also served as a Korean language interpreter.

After his honorable discharge, David built careers in software engineering, management consulting, and technology leadership, reaching senior positions at Raytheon, Honeywell, and Amazon Web Services. He holds a degree in Computer Science from Hawaii Pacific University and an MBA from University of Phoenix.

His other book, *The Real Money Guide*, combines personal finance instruction with lessons from his own journey through bankruptcy and recovery. He hosts the *Real Money Pathway* podcast and lives in the Nashville area with his wife Lacey.